T0202792

Human Chromosomes

Eeva Therman

Human Chromosomes
Structure, Behavior, Effects

Second Edition

With 87 Figures

Springer-Verlag
New York Berlin Heidelberg Tokyo

Eeva Therman
Laboratory of Genetics
University of Wisconsin
Madison, WI 53706
U.S.A.

Library of Congress Cataloging-in-Publication Data
Therman, Eeva.
    Human chromosomes.
    Includes bibliographies and indexes.
    1. Human chromosomes. 2. Human cytogenetics.
3. Human chromosome abnormalities. I. Title. [DNLM:
1. Chromosomes, Human. 2. Cytogenetics. QH 600 T411h]
QH431.T436 1985   611'.01816   85-17378

Media conversion by University Graphics Inc., Atlantic Highlands, New Jersey.

9 8 7 6 5 4 3 2 1

ISBN 978-0-387-96173-6   ISBN 978-1-4684-0269-8 (eBook)
DOI 10.1007/978-1-4684-0269-8

For Klaus Patau

Preface

"This book provides an introduction to human cytogenetics. It is also suitable for use as a text in a general cytogenetics course, since the basic features of chromosome structure and behavior are shared by all eukaryotes. Because my own background includes plant and animal cytogenetics, many of the examples are taken from organisms other than the human. Since the book is written from a cytogeneticist's point of view, human syndromes are described only as illustrations of the effects of abnormal chromosome constitutions on the phenotype. The selection of the phenomena to be discussed and of the photographs to illustrate them is, in many cases, subjective and arbitrary and is naturally influenced by my interests and the work done in our laboratory."

The above paragraph from the Preface of the first edition of this book also fits the present edition. However, so much has happened in five years in cytogenetics that—apart from a couple of pages here and there—the whole book has been rewritten and nine new chapters have been added.

The system used in the first edition to cite, whenever possible, the latest and/or the most comprehensive review rather than the original publications has been followed here also. Not only would complete literature citations increase the size of the book too much, but many readers have expressed satisfaction with the referencing method used here.

For suggestions and criticism I am greatly indebted to many colleagues, of whom I would especially like to mention Drs. Evelyn M. Kuhn, Richard Spritz, Millard Susman, and Bernard Weisblum. Furthermore, I appreciate the heroism of Drs. James F. Crow and Carter Denniston for reading and annotating the whole manuscript.

I am grateful to the cytogeneticists and editors who have generously permitted the use of published and unpublished pictures. I wish to express my special thanks to Dr. Carolyn Trunca for allowing me to use her unpublished analysis of the segregation of human translocations, and to Dr. Andrew Drewry for giving me unpublished data.

Without the help of members of my laboratory, Mrs. Barbara Susman and Mr. Walter Kugler, Jr., this book would hardly exist. Mrs. Susman has been involved in all phases from compiling literature lists to designing illustrations, and Mr. Kugler has done all the photographic work.

My sincere thanks are due to our secretaries, Ms. Juli-Kay Baumann and Mrs. Sandy Keller, who have typed and retyped the manuscript.

I wish to express my special gratitude to Dr. Ilse Riegel, whose help has been invaluable.

Madison, Wisconsin   Eeva Therman
November, 1985

Contents

I. Past and Future of Human Cytogenetics 1

The Past of Human Cytogenetics 1
The Dark Ages 2
The Hypotonic Era 2
The Trisomy Period.. 2
Chromosome Banding Era 3
Human Sex Chromosomes 4
Evolution of Human Chromosomes............................... 4
Nomenclature of Human Chromosomes 5
The Future of Human Cytogenetics 5
Ultrastructure of Eukaryotic Chromosomes 6
Longitudinal Differentiation of Chromosomes 6
Structure of Interphase Nuclei 7
Sex Determination and Sex Chromosomes 7
Mapping of Human Chromosomes 8
Chromosomes and Cancer.................................... 8
Somatic Cell Cytogenetics 8
Mutagenesis Studies.. 9
Chromosome Breakage Syndromes 9
Clinical Cytogenetics 10
References.. 10

II. Structure of the Eukaryotic Chromosome and the Karyotype 13

Metaphase Chromosome...................................... 13
Primary Constriction 13
Secondary Constrictions 15
Characterization of Metaphase Chromosomes 15
Chromosome Number.. 17
Chromosome Size 17
Shape of Chromosomes..................................... 18

DNA Content of Nuclei.. 18
Human Chromosome Complement 19
Banded Human Karyotype 21
Euploid Chromosome Changes............................... 21
Aneuploid Chromosome Changes........................... 23
Structural Changes in Chromosomes 24
References.. 24

III. Mitotic Cycle and Chromosome Reproduction............. 25

Significance of Mitosis.. 25
Interphase... 25
Prophase 27
Prometaphase 27
Metaphase... 27
Anaphase 27
Telophase 27
Chromosome Arrangement 27
Nondisjunction and Loss of Chromosomes 28
Mitotic Cycle 28
Chromosome Replication 29
References... 31

IV. Methods of Human Cytogenetics 32

Direct Methods 32
Tissue Culture Techniques 33
Prenatal Studies 33
Meiotic Studies.. 34
Sex Chromatin Techniques 35
Autoradiography 35
Banding Techniques of Fixed Chromosomes................. 36
Banding Techniques for Cells in Culture 37
Prophase Banding 38
Nomenclature of Human Chromosomes 38
Quantitative Methods 41
References... 42

V. Longitudinal Differentiation of Eukaryotic Chromosomes 45

Longitudinal Differentiation of Chromosomes 45
Molecular Differentiation of Chromosomes....................... 45
Prebanding Studies 46
Banding Studies on Human Chromosomes 47
Nucleoli and Chromocenters................................. 48
Constitutive Heterochromatin 50
Facultative Heterochromatin 52
Intercalary Heterochromatin 53
Chromosome Bands.. 53
Function of Human Chromosome Bands 54
References... 55

VI. Fine Structure and Function of the Eukaryotic Chromosome   58

Chromosome Fine Structure 58
Polytene and Lampbrush Chromosomes 61
References 63

VII. Chromosome Structural Aberrations 65

Origin of Structurally Abnormal Chromosomes 65
Chromosome Breaks and Rearrangements 66
Chromatid Breaks and Rearrangements 68
Telomeres 68
Telomere Association 70
The Origin of Dicentric Chromosomes Including Isodicentrics and
    Isochromosomes 70
Inactivation of the Centromere 71
Misdivision 72
Centric Fusion 72
Triradial and Multiradial Chromosomes 73
Fragile Regions 74
References 75

VIII. Causes of Chromosome Breaks 78

Spontaneous Chromosome Breaks 78
Radiation-Induced Breaks 79
Chemically Induced Breaks 80
Virus-Induced Breaks 80
Genetic Causes of Chromosome Breaks 82
Nonrandomness of Chromosome Breaks 82
Methods in Chromosome-Breakage Studies 83
Rules for Chromosome-Breakage Studies 84
References 85

IX. Chromosome Breakage Syndromes 88

Genotypic Chromosome Breakage 88
Bloom's Syndrome 88
Fanconi's Anemia 89
Ataxia Telangiectasia 91
Other Conditions with Increased Chromosome Aberrations 91
Cancer and Chromosome-Breakage Syndromes 92
References 92

X. Sister Chromatid Exchanges and Mitotic Crossing-Over 94

The Detection of Sister Chromatid Exchanges 94
The Occurrence of SCE 94
SCE in Mutagenesis Research 96
Significance of SCE 97
Mitotic Crossing-Over 97

Nonrandom Localization of Chiasmata   98
Segregation after Mitotic Recombination   99
The Origin of Mitotic Chiasmata   100
Gene Amplification   100
References   101

XI. Cell Fusion, Prematurely Condensed Chromosomes, and
     the Origin of Allocyclic Chromosomes   103

Cell Fusion   103
Prematurely Condensed Chromosomes (PCC)   103
Uses of PCC Formation   105
Origin of Allocyclic Chromosomes   107
References   108

XII. Modifications of Mitosis   110

Endoreduplication   110
Polyteny   112
Endomitosis   115
C-Mitosis   116
Restitution   116
Multipolar Mitoses   116
References   117

XIII. Somatic Cell Cytogenetics   119

History of Somatic Cell Cytogenetics   119
Methods of Somatic Cell Cytogenetics   119
Somatic Polyploidy   120
Structure of Nuclei in Somatic Cells   121
Amplification and Underreplication   122
Somatic Cell Cytogenetics and Differentiation   123
References   124

XIV. Main Features of Meiosis   125

Significance of Meiosis   125
Meiotic Stages   127
Premeiotic Interphase   128
Leptotene (or Leptonema)   128
Zygotene (or Zygonema)   128
Pachytene (or Pachynema)   132
Diplotene (or Diplonema)   133
Diakinesis   133
Metaphase I   133
Anaphase I   133
Telophase I   134
Interkinesis   134
Meiotic Division II   134

Some Meiotic Features 134
References.. 135

XV. Details of Meiosis 136

Structure of Chiasmata 136
Number of Chiasmata 137
Synaptonemal Complex.. 138
Meiotic Behavior of More Than Two Homologous Chromosomes 141
Human Meiosis 142
Premeiotic and Early Meiotic Stages in Man 143
Meiosis in Human Spermatocytes 144
Behavior of X and Y Chromosomes 144
References... 147

XVI. Meiotic Abnormalities 149

Nondisjunction of Autosomes.................................... 149
Nondisjunction of Sex Chromosomes 150
Misdivision of the Centromere 151
Chromosomally Abnormal Human Sperm 152
Male Infertility 153
Environmental Causes of Meiotic Nondisjunction 153
Maternal Age 154
Genetic Causes of Nondisjunction 156
Origin of Diploid Gametes 157
References... 157

XVII. Human Sex Determination and the Y Chromosome 160

The Y Chromosome 160
Sex Determination in Man 161
H-Y Antigen 162
Genes on the Y Chromosome................................... 163
Abnormal Y Chromosomes...................................... 164
References... 164

XVIII. Human X Chromosome 166

The Structure of the X Chromosome 166
X Chromatin or Barr Body Formation........................... 166
Inactivation of the X Chromosome 170
Reactivation of the X Chromosome.............................. 172
Sex Reversal and Intersexuality 172
References... 173

XIX. Numerical Sex Chromosome Abnormalities 176

Aneuploidy of X Chromosomes in Individuals with a Female
    Phenotype 176

Sex Chromosome Aneuploidy with Male Phenotype 179
Mosaicism 180
References... 180

XX. Structurally Abnormal X Chromosomes 182

Abnormal X Chromosomes Consisting of X Material 182
X Inactivation Center 185
X;Autosome Translocations and the Critical Region 187
Always-Active Regions on the Inactive X Chromosome 188
Translocations Involving the Y Chromosome..................... 188
Abnormal X Chromosome Constitutions and the Female Phenotype.... 189
Various Behavioral Abnormalities of the X Chromosome 190
Summary of Mammalian X Inactivation.......................... 191
References... 192

XXI. Numerically Abnormal Chromosome Constitutions in
Humans 194

Abnormalities of Human Chromosome Number 194
Polyploidy... 194
Human Triploids 195
Hydatidiform Moles 196
Human Tetraploids 196
Autosomal Aneuploidy 197
Anomalies Caused by Chromosomal Imbalance 198
13 Trisomy (D_1 Trisomy, Patau's Syndrome) 200
21 Trisomy Syndrome (Down's Syndrome, Mongolism) 203
18 Trisomy Syndrome (Edward's Syndrome) 203
Other Autosomal Aneuploidy Syndromes 204
Spontaneous Abortions 206
References... 207

XXII. Structurally Abnormal Human Autosomes................. 211

Structurally Abnormal Chromosomes 211
Chromosomal Polymorphisms 211
Quantitative Comparisons 213
Pericentric Inversions 213
Paracentric Inversions 215
Deletions or Partial Monosomies 215
Cri du Chat (Cat Cry) Syndrome 216
Ring Chromosomes 219
Insertions 222
Duplication or Pure Partial Trisomy 222
Dicentric Chromosomes 223
References... 225

XXIII. Reciprocal Translocations. 228

Occurrence 228
Breakpoints in Reciprocal Translocations 229

Multiple Rearrangements 230
Phenotypes of Balanced Translocation Carriers 230
Phenotypes of Unbalanced Translocation Carriers 231
Fetal Death 233
Examples of Translocation Families 234
Meiosis in Translocation Carriers 237
Genetic Risk for Translocation Carriers 238
References 240

XXIV. Robertsonian Translocations 243

Occurrence 243
Monocentric and Dicentric Chromosomes 244
Relative Frequencies of the Different Types of Robertsonian
    Translocations 246
Studies on the Newborn 247
Ascertainment Through an Unbalanced Individual 247
Ascertainment Through Infertility 247
Nonrandomness of Robertsonian Translocations 248
Segregation in Carriers of Robertsonian Translocations 250
Interchromosomal Effects 252
References 253

XXV. Double Minutes and Homogeneously Stained Regions ... 256

What are DMs and HSRs? 256
Structure of Double Minutes 256
Homogeneously Stained Regions and C-Minus Chromosomes 258
DMs and HSRs as Expressions of Gene Amplification 259
DMs and HSRs are Interchangeable 259
Origin of DMs and HSRs 260
Significance of HSRs and DMs 260
References 261

XXVI. Chromosomes and Oncogenes 262

What is Cancer? 262
Cancer Induction 263
Oncogenes 264
Reciprocal Translocations and Oncogenes 265
Recessive Oncogenes 270
Multistep Carcinogenesis 271
References 271

XXVII. Chromosomal Development of Cancer 273

Chromosomes and Cancer Progression 273
Chromosome Studies in Ascites Tumors 274
Chromosome Studies in Primary Tumors 274
The Apparent Predetermination of Chromosome Changes in Cancer ... 276
Mitotic Aberrations in Cancer Cells 276
Effect of Chromosome Changes on Tumor Development 278
References 280

XXVIII. Mapping of Human Chromosomes 282

Gene Mapping 282
Family Studies 283
Marker Chromosomes 283
Cell Hybridization 284
Selection 285
Chromosome Translocations 286
Other Uses of Deletions and Translocations in Mapping 287
Transfer of Microcells and Single Chromosomes 288
Gene Dosage 288
In-situ Hybridization 288
Recombinant DNA Gene Mapping Methods 290
Conclusions 291
References 291

Author Index 295

Subject Index 301

1
Past and Future of Human Cytogenetics

The Past of Human Cytogenetics

Before 1956 two "facts" were known about human cytogenetics. The human chromosome number was believed to be 48, and the XX–XY mechanism of sex determination was assumed to work in the same way as it does in *Drosophila*. Studies of the fruit fly had shown that the *ratio* of the number of X chromosomes to the number of sets of autosomes determines the sex of the organism. Both these fundamental notions about human chromosomes were eventually proved wrong.

The year 1956 is often given as the beginning of modern human cytogenetics, and indeed the discovery of Tjio and Levan (1956) that the human chromosome number is 46, instead of 48, was the starting point for subsequent spectacular developments in human chromosome studies. The difficulties of writing about even fairly recent history are well demonstrated by the very different accounts of this discovery related by the two participants themselves (Tjio, 1978; Levan, 1978).

The history of human cytogenetics has been reviewed at length several times, for instance, by Makino (1975) and more recently by Hsu (1979). Hsu's delightful book relieves this author of the responsibility of giving a detailed description of the developments in the field; instead I shall simply outline the major events in cytogenetics.

Hsu (1979) divides human cytogenetics conveniently into four eras: the dark ages before 1952, the hypotonic period from 1952 to about 1958, the trisomy period between 1959 and 1969, and the chromosome banding era that started in 1970 and still continues. To these should be added the prophase or high-resolution banding era, which has opened completely new avenues in human cytogenetics (Yunis, 1976). Finally there is the coming together on common ground of molecular and cytogenetic techniques from opposite directions. In the following discussion only a few highlights of these various developments will be recounted.

The Dark Ages

The difficulties faced by the early cytogeneticists are illustrated by a comparison of Fig. III.1 with the other photomicrographs of human chromosomes in this book. Despite the lack of clarity, the lymphocyte mitosis in Fig. III.1 shows the chromosomes considerably better than did the slides of paraffin-sectioned testes, stained with hematoxylin, that were used during the first quarter of this century. Of these studies only the paper by Painter (1923) is mentioned here, since it determined the ideas in this field for the next 33 years. Even though Painter's report that the human chromosome number was 48 was worded quite cautiously, the more often it was quoted, the more certain the finding seemed to become.

Despite the primitive techniques available, the groundwork for future studies was laid during these dark ages. The first satisfactory preparations of mammalian chromosomes were obtained by squashing ascites tumor cells of the mouse (Levan and Hauschka, 1952, 1953) and of the rat (cf. Makino, 1975). The first successful prefixation treatment with chemical substances was performed on mouse tumor cells by Bayreuther (1952); later, colchicine or its derivatives were used.

During this era, mammalian tissue culture techniques were developed. Prefixation treatment with hypotonic salt solution, which swells the cells and thus separates the chromosomes, was a decisive improvement in cytological technique. The hypotonic treatment was launched by Hsu (1952), although other laboratories were experimenting with similar treatments at the same time.

The Hypotonic Era

The simultaneous use of a number of new techniques made it possible to establish the correct chromosome number in man. They were the tissue culture and squash techniques combined with treatments with colchicine and hypotonic solution prior to fixation. Before the end of 1956, the finding of Tjio and Levan in embryonic lung cells was confirmed in human spermatocytes by Ford and Hamerton (1956), whose photomicrographs also showed that the X and Y chromosomes are attached end to end by their short arms in meiosis. During the hypotonic era the analysis of the human karyotype was also begun.

The Trisomy Period

The new techniques were soon applied to chromosome analyses of individuals who were mentally retarded or had other congenital anomalies, or both. The first autosomal trisomy was described by Lejeune et al

(1959), who found that mongolism (Down's syndrome) was caused by trisomy of one of the smallest human chromosomes. During the same year it was reported that Turner's syndrome was characterized by a 45,X chromosome constitution (Ford et al, 1959), and Klinefelter's syndrome by a 47,XXY chromosome complement (Jacobs and Strong, 1959). In addition, the first XXX woman was described (Jacobs et al, 1959). The observations on Turner's and Klinefelter's syndromes showed that the male sex in human beings is determined by the presence of the Y chromosome. Later it was established that the Y chromosome is effective in determining male sex even if it is combined with four X chromosomes; individuals with the XXXXY sex chromosome constitution are males, although abnormal.

The following year, D_1 trisomy (now known to be 13 trisomy) (Patau et al, 1960) and 18 trisomy (Edwards et al, 1960; Patau et al, 1960; Smith et al, 1960) were described. With these discoveries the viable autosomal trisomies—apart from mosaics—seemed to be exhausted, although the exceedingly rare 22 trisomy was later found, and chromosome studies turned to structural aberrations and their phenotypic consequences.

These developments coincided with an important innovation in cell culture technique. Nowell (1960) and Moorhead et al (1960) launched the short-term culture technique, using peripheral lymphocytes. The effectiveness of the technique was based on the mitosis-inducing ability of phytohemagglutinin. Such cultures, combined with the trick of drying the fixed chromosomes directly on microscope slides (Rothfels and Siminovitch, 1958), are still the most important source of human and mammalian chromosomes.

Chromosome Banding Era

Despite all claims to the contrary, the chromosomes in groups B, C, D, F, and G could not be identified individually on the basis of morphology (Patau, 1960); the numbers attached to the paired-off chromosomes in prebanding karyotypes represented sheer guesses. Although autoradiography had allowed the accurate identification of some chromosomes (cf. Patau, 1965), the degree of precision was increased by orders of magnitude with the introduction of chromosome banding techniques. In 1970, Caspersson et al applied fluorescence microscopy, which they had originally used to study plant chromosomes, to the analysis of the human karyotype. They discovered that the chromosomes consist of differentially fluorescent cross bands of various lengths. Careful study of these bands made possible the identification of all human chromosomes. This discovery was followed by a flood of different banding techniques that utilize either fluorescent dyes or the Giemsa stain. The banding of prophase chromosomes makes it possible to determine chromosome segments and breakpoints even more accurately (Yunis, 1976).

Another milestone was the discovery that chromosomes that incorporate bromodeoxyuridine (BrdU) instead of thymidine have different staining properties. This phenomenon has been successfully used to reveal the late-replicating chromosomes and chromosome segments (Latt, 1974). It also provides the basis for the study of sister chromatid exchanges (Latt, 1973).

It is much more difficult to obtain satisfactory chromosome preparations of the male meiosis—not to mention the female meiosis—in man than, for instance, in the mouse. But lately these difficulties have been overcome to some extent. The early stages of meiosis have been analyzed successfully in the oocytes (e.g., Therman and Sarto, 1977; Hultén et al, 1978), whereas work on the spermatocytes has yielded clear photomicrographs of the later stages (e.g., Stahl et al, 1973).

Human Sex Chromosomes

Throughout the four eras described by Hsu (1979), the understanding of the function and behavior of the mammalian sex chromosomes increased steadily. One of the first important observations was that the neural nuclei of the female cat had a condensed body, missing in the male nuclei (Barr and Bertram, 1949). This body has been called sex chromatin, the Barr body, or X chromatin.

The single active X hypothesis of Lyon (1961; cf. Russell, 1961) had a decisive influence on the entire field of mammalian sex chromosome studies. According to the Lyon hypothesis, as it is called, one X chromosome in mammalian female cells is inactivated at an early embryonic stage. The original choice of which X is inactivated is random, but in all the descendants of a particular cell the same X remains inactive. If a cell has more than two X chromosomes, all but one are turned off. This mechanism provides dosage compensation for X-linked genes because each cell, male or female, has only one X chromosome that is transcribed. The Barr body is formed by the inactive X chromosome (Ohno and Cattanach, 1962), which is out of step with the active X chromosome during the cell cycle.

One of the highlights in the study of mammalian sex determination is the recent discovery that the primary sex determination of the Y chromosome is mediated through the H–Y antigen, which is a plasma membrane protein (cf. Ohno, 1979). This antigen induces the development of testicular tissue, which in turn determines secondary sex development through its production of androgen.

Evolution of Human Chromosomes

The Phylogeny of Human Chromosomes, as Seuánez (1979) calls his book, has been studied intensively in recent years. A comparison of the chromosomes of man with those of his closest relatives—the chimpanzee, gorilla, and orangutan—shows that 99 percent of the chromosome bands are shared by the four genera. This similarity extends even to the prophase bands (Yunis and Prakash, 1982). The most prominent differences in banding patterns occur in the heterochromatic regions (cf. Seuánez, 1979).

The similarity of the chromosome banding pattern in all four genera demonstrates that most of the individual bands have retained their identity for more than 20 million years, and many of them for considerably longer. A number of chromosomes in man and the great apes are identical. The most conservative of these chromosomes is the X, which has not changed in morphology, at least between the monkey and man. Its gene content is assumed to have remained the same throughout mammalian development, or for some 125 million years (cf. Ohno, 1967; Seuánez, 1979). A comparison of the chromosomes of man and his relatives is now also under way on the molecular level (cf. Jones, 1977).

Nomenclature of Human Chromosomes

As the number of laboratories involved in the analysis of human chromosomes multiplied, so did the systems of chromosome designation. In an effort to create order in this threatening chaos, six conferences on chromosome nomenclature have been held: in Denver (1960), London (1963), Chicago (1966), and Paris (1971) (cf. Makino, 1975). The recommendations of the latest conferences in Stockholm and Paris (ISCN, 1978, 1981) include the designations of metaphase and prophase chromosome bands (Chapter IV).

The Future of Human Cytogenetics

The expansion of the science of human cytogenetics in somewhat less than 30 years is little short of miraculous; by now, from the viewpoint of cytogenetics, man is by far the most extensively studied organism. During its early stages, human cytogenetics was a more-or-less applied science: phenomena previously described in plants and animals were now being observed in man. However, human cytogenetics has come of age, and advances in this field have inspired studies in other branches of human genetics. Indeed it is the coordination of different approaches that has led to the most interesting results in this field. During the "dark ages", human cytogeneticists borrowed techniques from plant and animal studies. Now the opposite is often true, and both animal and plant chromosome studies owe a debt to the work done on humans.

How much has happened even during the last 5 years is obvious if one compares this book with its first edition (Therman, 1980). Cytogenetic studies have in many cases reached the point at which molecular biology takes over. However, without the basic cytogenetic knowledge, molecular work would not be possible.

Predictions of future developments in a scientific field can be based only on its present state. However, just as unexpected findings in the past have changed the course of events, they will undoubtedly do so in the future. In the following discussion, those approaches to human cytogenetics that seem most promising to this author are briefly reviewed.

Ultrastructure of Eukaryotic Chromosomes

The gap that existed for a long time between our knowledge of DNA and of chromosome structure on the light microscopic level is finally being bridged (Chapter VI). Although some of the steps still are hypothetical, we now have a fairly good idea how a several-centimeter-long DNA double helix is packed to form a few-micra-long chromatid. Interestingly, the structure as found in the giant lampbrush and polytene chromosomes seems in principle to apply to all eukaryotic chromosomes. Even gene action in lampbrush chromosomes is directly visible in the electron microscope.

In addition, "recombination nodules" are being studied in the electron microscope. These seem to determine the crossing-over sites in meiosis (Chapter XV). Obviously many of the mysteries surrounding chromosome pairing and crossing-over will be solved with similar studies. Actually much remains to be done on the early meiotic stages, even at the level of the light microscope.

Microspread and silver staining techniques have brought the synaptonemal complexes within reach of light microscopic examination offering interesting possibilities for the analysis of the pairing of structurally heterozygous chromosomes.

Longitudinal Differentiation of Chromosomes

The long-established fact that quinacrine-bright chromosome segments contain more heterochromatin, whereas the active genes are concentrated in the quinacrine-dark regions, is also reflected in the early replication of the latter and the late replication of the inactive chromosome segments during the S period.

Various banding and molecular studies have demonstrated the existence of different types of heterochromatin. However, the role—if any—that heterochromatin plays is still essentially unknown. This is also reflected in the variety of hypotheses, none of them backed up by solid evidence, concerning the possible effects of heterochromatin. These range from the idea that heterochromatin has no function, consisting of "selfish DNA", to the assumption that it has an important role in development and evolution.

The eukaryotic chromosomes contain large amounts of DNA which, though not heterochromatin, are not transcribed either. The possible role of this apparently inert chromatin is a challenge to future studies.

Although the so-called fragile regions have in recent years attracted a great deal of attention, both their structure and function are largely unknown. However, a connection has been claimed between these fragile sites and the constant chromosome breaks in cancer.

Now that the structure of telomeres and centromeres is being cleared up, similar studies should be done on the inactivated centromeres.

Structure of Interphase Nuclei

Although the arrangement of chromosomes in interphase nuclei attracted attention as early as the 1880s, only recently have techniques been developed that allow us to follow the chromosomes throughout the mitotic cycle (Chapters III and XI). One of these is cell fusion, which causes the condensation of the chromosomes in the interphase nucleus. Such studies have demonstrated that chromosomes occupy the same relative positions from one metaphase to the next. Apart from diptera and some other special cases, the arrangement of the chromosomes appears to be random.

Another useful tool in the study of interphase is the UV laser microbeam technique, which in interphase breaks a couple of neighboring chromosomes whose relative positions can be seen in the next metaphase.

Sex Determination and Sex Chromosomes

A factor(s) regulating the H–Y antigen is now known to lie on the short arm of the Y chromosome (Chapter XVII). Another important factor involved in sex determination is situated on the short arm of the X chromosome. The mode of action and relationship of these factors in the sex determination process are under intensive study.

The following features of the X chromosome are now generally accepted: the "critical region" on Xq, in which a break causes a position effect; the inactivation center on Xq and the always active region at the distal end of Xp are also firmly established. Under study is the possibility that other regions on the inactive X also stay active. The biochemical basis of neither X inactivation nor the phenomena just mentioned is known.

The BrdU technique has revealed that many exceptions to the previously accepted rules for X inactivation exist. These exceptions may to some extent help us to understand how abnormal X chromosome constitutions affect the phenotype.

The reactivation of a few genes on the inactive X chromosome has succeeded. Whether the absence of Barr bodies in many female cancers implies that two X chromosomes are active in their cells is still unclear.

Mapping of Human Chromosomes

The development of an accurate gene map is one of the main goals of the cytogenetic studies in any organism. The localization of genes to specific chromosomes and chromosome segments is one of the most rapidly advancing branches of human cytogenetics (Chapter XXVIII). The methodology includes linkage studies, the use of marker chromosomes in family studies, in vitro fusion of human cells with cells of other species, direct hybridization of DNA and RNA on chromosomes, and various molecular techniques. Previously it was possible to determine by hybridization the position of repeated genes, such as those coding for ribosomal RNA or histones. Now individual genes can also be mapped in the same way as soon as they are cloned.

Chromosomes and Cancer

The field of cytogenetics in which the most spectacular successes have been achieved during the last few years is the study of cancer chromosomes (Chapters XXV–XXVII). The old idea that, if the same chromosome aberration is consistently found in a certain type of malignant disease, it probably is its cause has finally proven itself. The observations on these constant chromosome aberrations, mostly reciprocal translocations, combined with virology and molecular biology, resulted in the discovery of oncogenes, and this finding has given an insight into the origin of cancer. Proto-oncogenes, which are present in all eukaryotic cells, can be activated through various processes, one of which is position effect. High-resolution banding has played a major role in the mapping of oncogenes. Although a unifying theory to explain these discoveries is still lacking, the field is advancing at tremendous speed, with new discoveries appearing almost weekly.

Homogeneously stained regions and double minutes, which consist of one or more genes amplified hundreds of times, are in some cases a step in carcinogenesis itself; in others they play a role in the progression of a tumor (Chapter XXV).

An interesting though still unanswered question is what causes the ubiquitous mitotic aberrations found in human cancers as well as in those of other mammals.

Somatic Cell Cytogenetics

A branch of mammalian cytogenetics that is just beginning is the analysis of what happens in the nuclei during development (Chapters XII–XIII). Important results in developmental cytology were achieved in plants and insects during the 1930s and 1940s. In mammals, however, mitotic modifications and their effects have been studied almost exclusively in liver, bone marrow, and malignant cells. Recently the cells of the normal trophoblast as well as of hydatidiform moles have also been analyzed. In addition to spectrophotometry, in use for 30 years, new techniques, such as flow cytometry, analysis of prematurely condensed chromosomes in cells fused at different stages of the cell cycle, and differential staining techniques are being applied to the analysis of interphase nuclei. Studies like these may shed light on the role that nuclear changes play in differentiation.

Amplification or underreplication of certain genes or whole chromosome segments seems also to be characteristic of differentiated cells.

Mutagenesis Studies

Chromosome breakage has been used for a long time as an indicator of the mutagenic effects of various agents (Chapters VIII–X). The resolution of such studies has been greatly increased by chromosome banding. Chromosome breaks occur nonrandomly along the chromosomes: the breaks are localized mainly in the Q-dark bands, some of which constitute veritable "hot spots".

The introduction of sister chromatid exchanges as a test system has caused a true revolution in mutagenesis testing. Sister chromatid exchanges are not only a much more sensitive indicator of mutagenic activity than chromosome breaks, but they are also considerably easier to score unambiguously.

However, in spite of these advances, trying to make sense of the mutagenesis literature is almost impossible. Apart from the main phenomena, for practically every claim there is a counterclaim, and a choice between them often seems impossible. If the vast amount of work in this important field is not to go to waste, standardization of the methodology is urgently needed.

A promising approach to the study of chromosome breaks before most of them have rejoined is provided by the cell fusion and premature chromosome condensation technique.

Chromosome Breakage Syndromes

Studies on the chromosome breakage syndromes have yielded inconsistent results (Chapters IX–X). For instance, it is still unclear whether chromosome damage decreases in Bloom's syndrome cells which are cultivated or fused with normal cells, or whether aberrations increase in the normal cells. The biochemical basis of these interesting syndromes is also unknown.

Bloom's syndrome shows a unique tendency to increased exchanges

10 I. Past and Future of Human Cytogenetics

between homologous chromosome segments. This allows us to study phenomena, such as mitotic crossing-over, which otherwise are very rare. Unequal exchanges apparently play an important role in gene amplification, in the variation of heterochromatic segments, and in the development of homogeneously stained regions and double minutes.

Clinical Cytogenetics

The vast amount of work done in clinical cytogenetics has led to the description of partial trisomy and monosomy syndromes for all chromosome arms (Chapters XXI–XXIV). Prophase banding, which has helped to determine the accurate breakpoints, has also resulted in the discovery of syndromes caused by minute deletions. This trend can safely be predicted to continue.

Chromosome analyses of defined populations have uncovered the causes of many birth defects and abortions. Examples of such groups are spontaneous abortions, newborn infants, mental retardates, infertile men, women with gonadal dysgenesis, and couples with repeated abortions; even the chromosome constitutions of individual sperms can now be determined.

More accurate empirical risk figures for different types of translocations and other chromosome aberrations are now emerging (Chapters XXIII–XXIV). Genetic counseling is further helped by the recent placental biopsy technique, which allows the diagnosis of chromosomally or biochemically abnormal fetuses at an early stage. Molecular techniques are also being applied to genetic counseling.

The following problems in clinical genetics still await solution. Do heterochromatic variants ever affect the phenotype or the nondisjunction rate of the chromosome concerned? Are the claimed interchromosomal effects of translocations, inversions, and heterochromatic variants real, or do they depend on ascertainment and publication bias? What role do position effects play in man, and are breaks in specific chromosome segments especially liable to affect the phenotype (Chapter XXIV)? Finally, are there genes in man that increase nondisjunction, and through what pathways do they act?

References

Barr ML, Bertram EG (1949) A morphological distinction between neurons of the male and female, and the behavior of the nuclear satellite during accelerated nucleoprotein synthesis. Nature 163: 676–677
Bayreuther K (1952) Der Chromosomenbestand des Ehrlich-Ascites-Tumors der Maus. Naturforsch 7: 554–557
Caspersson T, Zech L, Johansson C (1970) Differential banding of alkylating fluorochromes in human chromosomes. Exp Cell Res 60: 315–319

Edwards JH, Harnden DG, Cameron AH, et al (1960) A new trisomic syndrome. Lancet i: 787–790

Ford CE, Hamerton JL (1956) The chromosomes of man. Nature 178: 1020–1023

Ford CE, Jones KW, Polani PE, et al (1959) A sex-chromosome anomaly in a case of gonadal dysgenesis (Turner's syndrome). Lancet i: 711–713

Hsu TC (1952) Mammalian chromosomes in vitro. I. The karyotype of man. J Hered 43: 167–172

Hsu TC (1979) Human and mammalian cytogenetics. An historical perspective. Springer, Heidelberg

Hultén M, Luciani JM, Kirton V, et al (1978) The use and limitations of chiasma scoring with reference to human genetic mapping. Cytogenet Cell Genet 22: 37–58

ISCN (1978) An international system for human cytogenetic nomenclature. Birth defects: original article series, XIV:8. National Foundation, New York; also in Cytogenet Cell Genet 21: 309–404

ISCN (1981) An international system for human cytogenetic nomenclature—high resolution banding. Birth defects: original article series, XVII:5. National Foundation, New York; also in Cytogenet Cell Genet 31: 1–84

Jacobs PA, Baikie AG, Court Brown WM, et al (1959) Evidence for the existence of the human "super female." Lancet ii: 423–425

Jacobs PA, Strong JA (1959) A case of human intersexuality having a possible XXY sex-determining mechanism. Nature 183: 302–303

Jones KW (1977) Repetitive DNA and primate evolution. In: Yunis JJ (ed) Molecular structure of human chromosomes. Academic, New York; pp 295–326

Latt SA (1973) Microfluorometric detection of deoxyribonucleic acid replication in human metaphase chromosomes. Proc Natl Acad Sci USA 70: 3395–3399

Latt SA (1974) Microfluorometric analysis of DNA replication in human X chromosomes. Exp Cell Res 86: 412–415

Lejeune J, Gautier M, Turpin R (1959) Etude des chromosomes somatiques de neuf enfants mongoliens. Compt Rend 248: 1721–1722

Levan A (1978) The background to the determination of the human chromosome number. Am J Obstet Gynecol 130: 725–726

Levan A, Hauschka TS (1952) Chromosome numbers of three mouse ascites tumors. Hereditas 38: 251–255

Levan A, Hauschka TS (1953) Endomitotic reduplication mechanisms in ascites tumors of the mouse. J Natl Cancer Inst 14: 1–43

Lyon MF (1961) Gene action in the X-chromosome of the mouse. Nature 190: 372–373

Makino S (1975) Human chromosomes. Igaku Shoin, Tokyo

Moorhead PS, Nowell PC, Mellman WJ, et al (1960) Chromosome preparations of leucocytes cultured from human peripheral blood. Exp Cell Res 20: 613–616

Nowell PC (1960) Phytohemagglutinin: an initiator of mitosis in cultures of normal human leukocytes. Cancer Res 20: 462–466

Ohno S (1967) Sex chromosomes and sex-linked genes. Springer, Heidelberg

Ohno S (1979) Major sex-determining genes. Springer, Heidelberg

Ohno S, Cattanach BM (1962) Cytological study of an X-autosome translocation in *Mus musculus*. Cytogenetics 1: 129–140

Painter TS (1923) Studies in mammalian spermatogenesis. II. The spermatogenesis of man. J Exp Zool 37: 291–336

Paris Conference (1971) Standardization in human cytogenetics. Birth defects: original article series, VIII:7. New York: The National Foundation, 1972

Patau K (1960) The identification of individual chromosomes, especially in man. Am J Hum Genet 12: 250–276

Patau K (1965) Identification of chromosomes. In: Yunis JJ (ed) Human chromosome methodology. Academic, New York; pp 155–186

Patau K, Smith DW, Therman E, et al (1960) Multiple congenital anomaly caused by an extra autosome. Lancet i: 790–793

Rothfels KH, Siminovitch L (1958) An air-drying technique for flattening chromosomes in mammalian cells grown in vitro. Stain Tech 33: 73–77

Russell LB (1961) Genetics of mammalian sex chromosomes. Science 133: 1795-1803

Seuánez HN (1979) The phylogeny of human chromosomes. Springer, Berlin

Smith DW, Patau K, Therman E, et al (1960) A new autosomal trisomy syndrome: multiple congenital anomalies caused by an extra chromosome. J Pediatr 57: 338–345

Stahl A, Luciani JM, Devictor-Vuillet M (1973) Etude chromosomique de la meiose. In: Boué A, Thibault C (eds) Les accidents chromosomiques de la reproduction. Paris: I.N.S.E.R.M., Centre International de l'Enfance, pp 197–218

Therman E (1980) Human chromosomes. Springer, New York

Therman E, Sarto GE (1977) Premeiotic and early meiotic stages in the pollen mother cells of *Eremurus* and in human embryonic oocytes. Hum Genet 35: 137–151

Tjio JH (1978) The chromosome number of man. Am J Obstet Gynecol 130: 723–724

Tjio JH, Levan A (1956) The chromosome number in man. Hereditas 42: 1–6

Yunis JJ (1976) High resolution of human chromosomes. Science 191: 1268–1270

Yunis JJ, Prakash O (1982) The origin of man: A chromosomal pictorial legacy. Science 215: 1525–1530

II
Structure of the Eukaryotic Chromosome and the Karyotype

Metaphase Chromosome

Higher organisms are *eukaryotes* in contrast to bacteria and phages, which are *prokaryotes*. The eukaryote chromosome is a complicated structure that, in addition to DNA, contains several different types of proteins. The chromosomes of higher organisms are studied most frequently at mitotic metaphase. This is the stage at which the chromosomes reach their greatest condensation, and this natural condition is increased by a prefixation treatment with various drugs, for example, colchicine. During mitotic metaphase, the condensed chromosomes appear in identifiable shapes characteristic of the karyotype of the species being studied.

Primary Constriction

A typical metaphase chromosome consists of two arms separated by a *primary constriction*, which is made more clearly visible by treatment with colchicine. This constriction marks the location of the *centromere* or *spindle attachment* (sometimes called the kinetochore), which is essential for the normal movements of the chromosomes in relation to the spindle. A chromosome without a centromere is an *acentric* fragment and either is lost or drifts passively when the other chromatids move to the poles, by the action of the spindle, during anaphase.

In the plant genera of rushes (*Juncus*) and sedges (*Carex*), as well as in certain insects and the scorpions, the centromere is diffuse or multiple, and the chromosomes lack a primary constriction. This means that a small chromosomal fragment, even when separated from the rest of the chromosome, acts like a complete chromosome and displays normal anaphase movements.

A metaphase chromosome consists of two *sister chromatids* that separate in mitotic anaphase. The genetic constituent of a chromatid is a double helix of DNA. Cytologists have long disputed whether a chromatid consists of one double helix of DNA or of several parallel ones. It was formerly assumed that the latter situation prevails, especially in organisms with large chromosomes. However, recent observations show convincingly that each chromatid of a eukaryotic chromosome contains one double helix of DNA continuous from one end of the chromosome to the other. In other words, a chromatid is a *unineme* structure. Certain specialized tissues, such as the endosperm in plant seeds wherein the chromatids consist of more than one DNA strand, provide exceptions to this rule. How widespread such *multineme* chromosomes are is not clear at present.

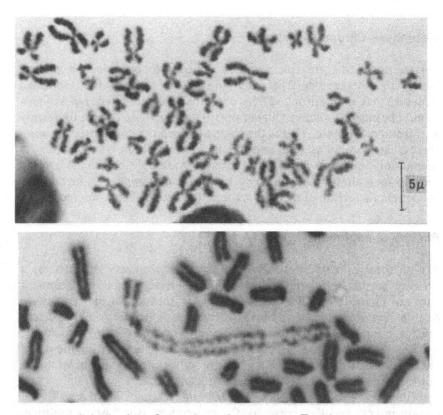

Figure II.1. Spiralization of metaphase chromosomes. Top: human lymphocyte. Bottom: mouse cancer cell treated with 1-methyl-2-benzylhydrazine (Therman, 1972).

Through a series of coils within coils, the chromosome strands, or *chromonemata*, shorten greatly between interphase and metaphase. The largest coil is sometimes visible in the light microscope, as in a mouse cancer chromosome treated with 1-methyl–2-benzylhydrazine (Fig. II.1), which displays segments with distinct coiling interspersed with condensed regions (Therman, 1972). In addition to DNA, eukaryotic chromosomes contain nonhistone proteins together with five types of histones (cf. Ris and Korenberg, 1979), all of which seem to play a role in the condensation and coiling of chromosomes.

Secondary Constrictions

A chromosome may also contain a *secondary constriction*, which appears as an unstained gap in the chromosome. Usually a secondary constriction contains a *nucleolar organizer* (Fig. II.2), and may be situated anywhere along the chromosome. However, these secondary constrictions are most often near an end, separating a small segment, a *satellite*, from the main body of the chromosome. In such cases the secondary constriction is called a *satellite stalk*.

Characterization of Metaphase Chromosomes

A metaphase chromosome is identified morphologically both by its total length and by the position of the centromere, which determines the relative lengths of its arms. A secondary constriction also helps in the identification of a particular chromosome. Nowadays chromosomes are usually identified by means of banding techniques (Chapters IV and V).

A chromosome in which the centromere is near the middle is called *metacentric*. In an *acrocentric* chromosome, the arms are markedly unequal. Chromosomes intermediate between these two are *submetacentric* and *subtelocentric*. A chromosome in which the centromere is at the very end is *telocentric*. Truly telocentric chromosomes are almost nonexistent in natural populations, but they have been found in some unusual individuals, including man.

In human chromosomes, the short arm is designated p (petite) and the long arm q (the next letter in the alphabet). Chromosomes are usually characterized by one of two parameters. The *arm ratio* (q/p) is the length of the long arm divided by that of the short arm. The *centromere index* expresses the percentage of the short arm in terms of the total chromosome length $[(p/p+q) \times 100]$. The length of a particular chromosome relative to others in the same plate, together with the arm ratio or centromere index, is sometimes sufficient to permit identification of the chromosome.

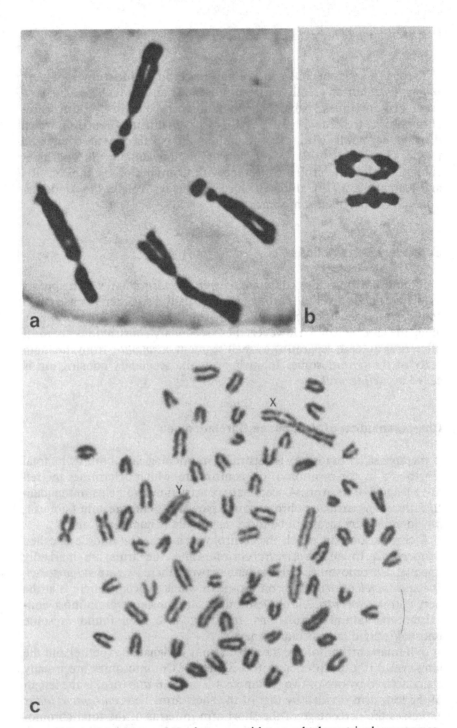

Figure II.2. (a) Metaphase of *Haplopappus* with two subtelocentric chromosomes (nucleolar constrictions in the short arms) and two submetacentric chromosomes; (b) two bivalents in I meiotic metaphase of *Haplopappus* (a and b courtesy of RC Jackson); (c) metaphase of male reindeer (*2n* = 76) showing X and Y chromosomes.

Chromosome Number

The diploid chromosome number of an organism, usually determined by counting the chromosomes in dividing somatic cells, is indicated by the symbol $2n$. The gametes have one-half the diploid number (a haploid set), indicated by n.

The chromosome numbers vary greatly between and within groups of organisms. The chromosome numbers show no clear trend of becoming either higher or lower during evolution. A haploid number of $n = 2$ is extremely rare, the best known example being the compositous plant, *Haplopappus* (Fig. II.2), which grows in the southwestern United States. In mammals, the lowest haploid number $n = 3$ has been found in the female muntjac, a small Indian deer (the male muntjac has $2n = 7$ chromosomes). The largest haploid number reported in a higher organism is about $n = 630$ in a fern (*Ophioglossum reticulatum*). In mammals the highest haploid number $n = 46$ has been observed in a rodent (*Anotomys leander*). The chromosome complements established for mammalian species have been reviewed by Hsu and Benirschke (1967–1977).

The haploid number of most organisms is between 6 and 25. In Fig. II.2 two widely different chromosome constitutions are illustrated. In *Haplopappus* ($2n = 4$) one pair of chromosomes is submetacentric; the other is subtelocentric and has a nucleolar constriction in the short arm. The reindeer has a relatively high chromosome number for a mammal ($2n = 76$). Apart from a few subtelocentric chromosome pairs, most reindeer chromosomes are acrocentric. The metacentric X and the acrocentric Y are clearly distinguishable from the autosomes.

Chromosome Size

Chromosome size also varies widely in different organisms, ranging from a fraction of a micron (μ) in length, which is at the limit of resolution of the light microscope, to more than 30 μ. Very small chromosomes are found in the fungi and green algae, whereas the largest ones have been observed in some amphibians and liliaceous plants. Most grasshoppers also have large chromosomes.

Although the chromosome complements of different organisms tend to be more similar the more closely related they are, there are striking exceptions to this rule. One of the classic examples of a great size difference in the chromosomes of two related species with the same diploid number ($2n = 12$) is provided by the leguminous plant *Lotus tenuis*, in which the mean length of the chromosomes is 1.8 μ, and *Vicia faba*, in which the corresponding value is 14.0 μ (cf. Stebbins, 1971).

In general, higher organisms tend to have larger chromosomes than do lower ones. Exceptions to this rule are also numerous, however, both among unicellular and multicellular organisms.

The chromosomes within the same chromosome complement usually fall into a fairly limited size range; in other words, they all tend to be either large or small. Yet in a few instances in some animal groups, such as birds and lizards, the chromosome complements consist of a number of large chromosomes and a larger number of very small microchromosomes. In metaphase the small chromosomes usually lie in the middle of the plate, with the large ones forming a circle around them.

Shape of Chromosomes

In addition to the number and size of the chromosomes, the chromosome complement of a species is characterized by the shape of the chromosomes. They may all be of one type or a combination of different types. In the mouse, for instance, all the chromosomes are acrocentric (Fig. VIII.1); in man the chromosomes range from metacentric to acrocentric (Figs. II.1 and II.3). Within a group, the more highly developed species tend to have more asymmetrical (the two arms are unequal) chromosomes, which have evolved from species with more metacentric complements (cf. Stebbins, 1971).

DNA Content of Nuclei

The DNA content of a nucleus is determined by the number and size of the chromosomes of the organism. In the animal kingdom, the values range from 168.0 picogram (pg) per diploid nucleus in the salamander

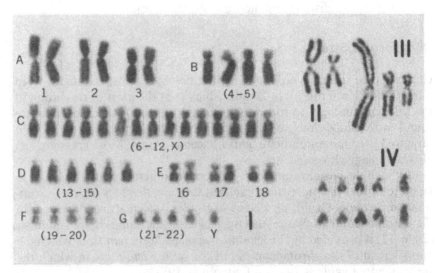

Figure II.3. (I) Normal human male karyotype from a lymphocyte. (II) Chromosomes 1 and 9 showing fuzzy regions, which now are known to represent heterochromatin. (III) Chromosomes 1, 9, and 16 showing fuzzy regions. (IV) G and Y chromosomes from father and son (orcein staining).

Amphiuma, which has very large chromosomes, to 0.2 pg in Drosophila, a 1000-fold difference. In plants the differences are almost as great, with values ranging from 196.7 pg in the liliaceous plant *Fritillaria* to 1.4 pg in the flax (*Linum usitatissimum*) (cf. Rees and Jones, 1977). In man, the DNA content of a diploid nucleus is 6.4 pg (cf. Rees and Jones, 1977). The values in other mammals deviate surprisingly little from this, especially when we consider the variation shown by their chromosome numbers and sizes.

Human Chromosome Complement

Man has 44 autosomes (nonsex chromosomes) and two sex chromosomes (two X chromosomes in the female and one X and one Y chromosome in the male). Human chromosomes range in size from somewhat larger than 5 μ to less than 1 μ; the range, however, varies between cells. With respect to both chromosome number and size, man stands in the middle range of higher organisms.

The term *karyotype* describes a display of the chromosomes of an organism in which they are lined up, starting with the largest and with shorter chromosome arm pointing to the top of the page. An *idiogram* is a diagrammatic karyotype based on chromosome measurements in many cells. Figure II.3 illustrates the karyotype of a normal human male. Morphological identification is based on the relative sizes of the chromosomes and their arm ratios. According to these criteria, chromosomes 1, 2, 3, 16, 17, 18, and the Y can be individually distinguished. Chromosome 9 (C′ in prebanding karyotypes) sometimes shows a fuzzy region next to the centromere on the long arm and can be identified on that basis. The rest of the chromosomes can be classified only as belonging to the groups B, C (which contains the X chromosome), D, F, and G. Chromosomes 1 and 3 are typically metacentric; chromosome 2 is on the borderline between metacentric and submetacentric; the B chromosomes represent the subtelocentric type. The D and G chromosomes are acrocentric and usually display satellites on the short arms. The sizes of both short arms and satellites vary in different persons. Sometimes satellites are so small that they are practically invisible in the light microscope.

With the banding techniques now available, all the human chromosomes can be identified. However, even in the prebanding era, attempts were made to distinguish the chromosomes by measuring them and pairing off those that were most similar in size. However, even identifiable homologous chromosomes, e.g., two number 1 chromosomes in the same metaphase plate, may show considerable difference in length. The coefficient of variation may be 5 percent or more. Since the real size difference between nonhomologous B, C, and D chromosomes is *less than* 5 percent (see Table II.1), pairing off by measurement alone clearly cannot lead to the correct identification of chromosomes, except sometimes by chance

Table II.1. Relative Lengths of Human Chromosomes[a]

Chromosome		Average length (in % of autosomal genome) of:		Chromosome		Average length (in % of autosomal genome) of:	
Group	No.	Long arm	Short arm	Group	No.	Long arm	Short arm
A	1	4.68	4.57		13	3.29	—
	2	5.28	3.35	D	14	3.12	—
	3	3.80	3.32		15	2.89	—
B	4	4.85	1.84		16	1.93	1.34
	5	4.66	1.75	E	17	2.07	0.96
	6	3.87	2.36		18	2.04	0.76
	7	3.54	2.04	F	19	1.32	1.11
	8	3.45	1.63		20	1.30	1.05
C	9	3.23	1.72	G	21	1.26	—
	10	3.22	1.54		22	1.38	—
	11	2.90	1.88	Total autosomes		100.00	
	12	3.38	1.32				
					X	3.26	2.02
					Y	1.64	—

[a]Revised by Patau (unpublished).
— Short arms not measurable.

(Patau, 1965). Despite this fact, attempts have been made to identify the X chromosome by pairing off the C chromosomes in a male. Those who started with the longest pair claimed that X was the shortest C chromosome; and those who started from the shortest chromosomes found that the partnerless X was the largest of the C group chromosomes. In reality it is the third largest (Table II.1). Obviously the pairing-off method has not been useful. It is equally useless for identifying members of the other chromosome groups. Consequently most karyotype numbers in prebanding days represent sheer guesses and should be so interpreted (cf. Patau, 1965).

Autoradiography, which is based on the finding that different chromosomes and chromosome segments replicate at different times during the synthetic period in interphase, provided a step forward in chromosome identification. This technique allows the observer to identify the inactive X chromosome in the female. The phenomenon of X inactivation will be discussed in Chapter XVIII. Also, chromosomes 4 and 5 can be distinguished, as can 13, 14, and 15, and 21 and 22 (cf. Patau, 1965).

However, autoradiography is a laborious process and has been superseded for chromosome identification by the banding techniques.

Banded Human Karyotype

Although the banding techniques and their use are discussed in detail in Chapters IV and V, a G-banded human karyotype is presented here (Fig. II.4). The chromosomes are numbered according to Caspersson et al (1971; see also Paris Conference, 1971), who first published a banded human karyotype. The chromosomes are arranged numerically according to length, with one exception: chromosome 22 is actually longer than 21. Since the chromosome that, in the trisomic state, causes mongolism (Down's syndrome) has long been called 21 in the literature, it was thought impractical and confusing to reverse the numbers. Chromosome 21 is, therefore, defined by the syndrome it causes. As seen in Fig. II.4, each human chromosome can be distinguished by its banding pattern. Most of the individual arms can also be unambiguously identified.

In Table II.1 the lengths of human chromosome arms are expressed as a percentage of the length of the haploid autosomal set or genome. The values are averages based on the compilation of a large number of measurements by different authors, with certain systematic errors corrected (Patau, unpublished). The lengths of the short arms of the acrocentric chromosomes are not given, both because they vary between individuals and because anything under 1 μ (which would be about 1 percent of the haploid karyotype) is only about twice the wavelength of visible light; therefore the measurement of the much smaller short arms would be of dubious value.

Euploid Chromosome Changes

An interesting question is: How have the widely different karyotypes diverged from each other during evolution? Evolution is based on changes that occur in the genetic material upon which natural selection acts. The genetic material may undergo qualitative gene changes, i.e., *gene mutations*. However, the evolution of karyotypes depends mainly on changes in the quantity or arrangement of the genes. These larger changes constitute the so-called *chromosome mutations*.

Changes in the number of whole sets of chromosomes, or *euploid* changes, usually lead to *polyploidy*. *Haploidy*, as a rule, is not a viable condition. Polyploidy can be inferred when the chromosome number of an organism is a multiple of the haploid chromosome number of a related species. A chromosome complement containing three haploid sets is called *triploid* (*3n*). Four sets constitute a *tetraploid* chromosome complement (*4n*). It is estimated that about one-half of higher plants have chromosome numbers that are multiples of those of other related species.

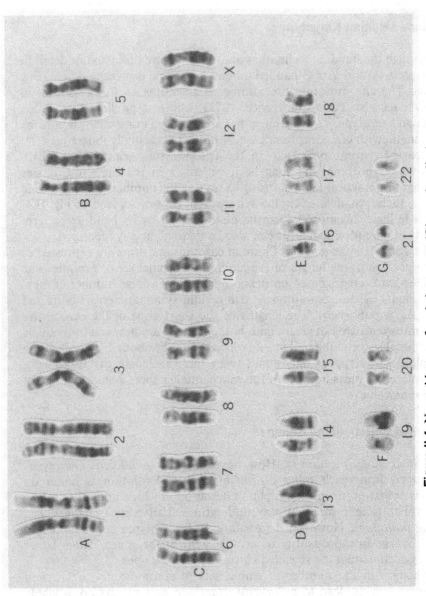

Figure II.4. Normal human female karyotype (Giemsa-banding).

In certain plant families, such as grasses (*Graminae*) or roses (*Rosaceae*), three-fourths of the species are polyploid. For example, in the genus *Rubus* of the rose family, the following multiples of the basic number 7 are found: 14, 21, 28, 35, 42, 49, 56, 63, and 84. In the compositous genus *Chrysanthemum* ($n = 9$), the diploid numbers range from 18 to 198 (cf. Darlington and Wylie, 1955).

In plant evolution, polyploidy is often combined with hybridization between different species. When the chromosome number of a hybrid is duplicated, we may see in one step the emergence of a new species combining the diploid complements of the parents. A well-known cultivated plant of this type is *Triticale*, a hybrid of wheat and rye with the chromosome complements of both species ($2n = 56$). However, one of the parents, the cultivated wheat, is a hexaploid with six sets of 7 chromosomes (one of the basic numbers in the grasses). Rye, on the other hand, is a diploid with $2n = 14$ chromosomes.

Polyploidy, which has been very important in plant evolution, seems to have played almost no role in animal evolution. However, in fishes and lizards there are indications, and in amphibians definite proof, of polyploidy (Beçak et al, 1967). Polyploidy in animals upsets the sex determination mechanism, and this has generally been assumed to block the successful establishment of polyploidy in them. Indeed, most polyploid animal species are *asexual*; they produce offspring *without* meiosis or fertilization. As a rule, such offspring are uniform and genetically identical with the parent. Because asexual reproduction takes place without meiosis or fertilization, the results of the process naturally lack the variability created by meiotic crossing-over and segregation.

From an evolutionary point of view, asexual species have reached a dead end, no matter how successful they may be under specific prevailing conditions. Polyploid parthenogenetic forms are found in shrimp, earthworms, and some insects (cf. Darlington, 1958). In plants, asexual reproduction through rhizomes, bulbs, runners, and other organs is common, especially in polyploid species. Apomictic asexual seed formation corresponds to parthenogenesis in animals. However, all organisms reproducing asexually can be said to have sold their future for a present advantage (cf. Darlington, 1958).

Aneuploid Chromosome Changes

Chromosome mutations also include changes in the number of individual chromosomes, as opposed to sets of chromosomes. These changes in the number of normal chromosomes, leading to unequal numbers of different chromosomes, are referred to as *aneuploidy*. The absence of one chromosome from the diploid complement is *monosomy*; the presence of an extra one is *trisomy*. A chromosome complement with two extra identical chromosomes is *tetrasomic*.

Stuctural Changes in Chromosomes

The third type of chromosome mutation occurs as a result of chromosome breakage and rejoining in such a way that the chromosomes are structurally reorganized. The details of chromosome rearrangements are discussed in Chapter VII. Such changes may sometimes result in an increase or decrease in the number of chromosome segments as well as in changes of their arrangement.

Various changes in chromosome number and structure have created the multitude of karyotypes observed in plants and animals. However, whatever selective advantage is bestowed on a given species by chromosome changes—few or numerous, large or small, symmetric or asymmetric—is a much discussed but still unresolved question (cf. Stebbins, 1971).

References

Beçak ML, Beçak W, Rabello MN (1967) Further studies on polyploid amphibians (*Ceratophrydidae*). I. Mitotic and meiotic aspects. Chromosoma 22: 192–201

Caspersson T, Lomakka G, Zech L (1971) 24 fluorescence patterns of human metaphase chromosomes—distinguishing characters and variability. Hereditas 67: 89–102

Darlington CD (1958) The evolution of genetic systems. Basic Books, New York

Darlington CD, Wylie AP (1955) Chromosome atlas of flowering plants. Hafner, New York

Hsu TC, Benirschke K (1967–1977) An atlas of mammalian chromosomes. Springer, Heidelberg, Vol 1–10

Paris Conference (1971) Standardization in human cytogenetics. Birth defects: original article series, VIII: 7. New York: The National Foundation, 1972

Patau K (1965) Identification of chromosomes. In: Yunis JJ (ed) Human chromosome methodology. Academic, New York, pp 155–186

Rees H, Jones RN (1977) Chromosome genetics. University Park Press, Baltimore

Ris H, Korenberg JR (1979) Chromosome structure and levels of chromosome organization. In: Goldstein L, Prescott DM (eds) Cell biology. Academic, New York, pp 268–361

Stebbins GL (1971) Chromosomal evolution in higher plants. Addison-Wesley, Menlo Park, California

Therman E (1972) Chromosome breakage by 1-methyl-2-benzylhydrazine in mouse cancer cells. Cancer Res 32: 1133–1136

III
Mitotic Cycle and Chromosome Reproduction

Significance of Mitosis

In mitosis the genetic material of a cell is divided equally and exactly between two daughter cells. Each chromosome replicates in interphase, then divides into two daughter chromosomes, which segregate in anaphase and, with the other chromosomes of the set, form two daughter nuclei. In other words, the cells undergo a regular alternation of chromosome replication and segregation. The main features of mitosis are strikingly similar and may be observed throughout eukaryotes, from unicellular organisms to man. Figure III.1 illustrates mitosis in untreated human lymphocytes. (Nowadays, only a few cytogeneticists have even seen human chromosomes that have not been treated with colchicine and hypotonic solution.) For comparison, Fig. XIV.1 shows the last premeiotic mitosis in untreated pollen mother cells of the liliaceous plant, *Eremurus* ($2n = 14$).

Interphase

The mitotic cycle consists of the mitotic stages and the interphase. In interphase the chromosomes are long, despiralized threads that are individually indistinguishable. Interphase nuclei come in diverse shapes from nearly spherical or oblong to lobed or branched structures. These last two forms are exceptional but have been seen, for instance, in many insects. In the light microscope, interphase nuclei appear more or less evenly stained with certain condensed chromosome segments, the so-called chromocenters, which stand out. The shapes and sizes of chromocenters

vary greatly, not only among species but also in different tissues of the
same organism (Fig. V.2.c–e). One or more nucleoli are usually visible in
an interphase nucleus (Fig. V.2a). Since the nucleoli tend to fuse, the largest number observed reflects the true number of active nucleolar organizers in the organism.

An interphase nucleus has often been described as the resting nucleus,

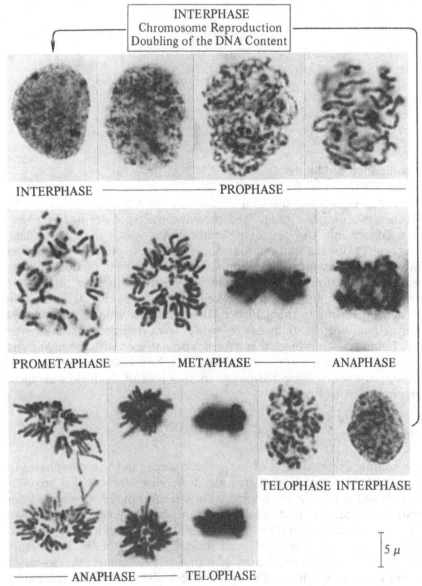

Figure III.1. Mitotic cycle in cultured human lymphocytes (Feulgen squash).

a singularly inappropriate term, since most of the biochemical activity of the nucleus takes place during this phase. Perhaps interphase nuclei in differentiated tissues "rest" more than the interphases in cells that are still dividing.

Prophase

In prophase, the chromosomes first become visible as thin long threads that gradually shorten and thicken as the diameter of the chromosome coil increases (Figs. III.1 and XIV.1). At the same time that the threads are shortening, the nucleoli vanish.

Prometaphase

Prophase is followed by a short prometaphase. During this period the nuclear membrane dissolves and the chromosomes, which are nearing their maximum condensation, collect on a metaphase plate.

Metaphase

Outside the nucleus, an organelle called the *centriole* or *centrosome* has divided and the mitotic spindle develops between the centrioles. The centromeres of the chromosomes collect halfway between the poles (centrioles) to form a metaphase plate (Fig. III.1). Long chromosome arms may stick out of the plate. Even though plants do not possess defined centrioles, the spindle arises between two polar areas.

Anaphase

The centromeres divide and the spindle fibers drag the sister chromatids to opposite poles (Figs. III.1 and XIV.1).

Telophase

Nuclear membranes are formed around the two chromosome groups. Gradually the chromosome coils loosen, the individual chromosomes become indistinguishable (Figs. III.1 and XIV.1), and the nucleoli reappear. Telophase is usually followed by cytoplasmic division, after which the nuclei revert once more to interphase.

Chromosome Arrangement

Interphase chromosomes usually show a so-called *Rabl orientation*, named after the German cytologist Carl Rabl, who in 1885 described this phenomenon (cf. Sperling, 1982). In a Rabl orientation the centromeres lie polarized in the position they reached in the previous anaphase. In prophase the chromosomes reappear in the same position (Sperling and Lüdtke, 1981).

Apart from the smaller chromosomes, which usually lie in the middle of a metaphase plate, and the association of satellites and heterochromatic regions, in most organisms the relative positions of the chromosomes in interphase and the following metaphase seem to be random (cf. Therman and Denniston, 1984). The most notable exceptions are the dipteran insects, in which the homologous chromosomes are paired throughout the mitotic cycle.

Although claims of nonrandom chromosome arrangements have been put forward, as a rule they are not borne out by solid evidence. On the other hand, convincing evidence for a random chromosome arrangement has been provided by cell fusion, which resulted in prematurely condensed chromosomes (Chapter XI), combined with a UV laser microbeam technique (Sperling and Lüdtke, 1981; Cremer et al, 1982).

Nondisjunction and Loss of Chromosomes

For various reasons the orderly segregation of daughter chromosomes in anaphase may sometimes fail. Like other biological processes, it is more likely to go wrong in older persons. The inclusion of *both* daughter chromosomes in the *same* nucleus, whatever the mechanism or reason, is called *nondisjunction*. As a result, one daughter cell gains an extra chromosome, becoming trisomic, whereas the other loses a chromosome and ends up being monosomic.

One daughter chromosome—or sometimes both—may lag behind in its division and not reach either pole. Such laggard chromosomes form micronuclei in interphase. These nuclei usually do not divide, and consequently the chromosomes included in them are lost from the complements of both daughter cells.

If the trisomic and monosomic cells arising in somatic tissues through nondisjunction or chromosome loss are viable, the result is mosaicism. Mosaicism may occur in a tissue or, if nondisjunction takes place very early in development, the entire organism may be mosaic. Actually we are all mosaics to some degree.

Mitotic Cycle

The duration of the mitotic cycle varies greatly in different organisms and in different tissues. The cleavage divisions of a toad take 15 minutes, whereas the same process in mouse ear epidermis lasts more than 40 hours (cf. White, 1973). The general rule seems to be that the larger the amount of nuclear DNA, the longer the duration of the mitotic cycle. Thus, organisms that are polyploid or have an otherwise high nuclear DNA content display the longest cycles.

The relative lengths of the individual phases also vary, although less than the variability in the duration of the whole cycle. Interphase usually lasts much longer than mitosis. Of the interphasic stages, Gap 1(G_1), which lies between the end of telophase and the beginning of the Synthesis (S) periods, is the longest; DNA synthesis (chromosome replication) takes place during the S period. This stage is succeeded by Gap 2 (G_2), which lasts until prophase. A chromosome is made up of one chromatid during G_1 and two in the G_2 phase.

As already mentioned, the number of haploid chromosome sets in a cell is indicated by the symbols *n*, *2n*, *3n*, and so on. The symbol for the amount of DNA, as opposed to the number of chromosome sets, is C. The haploid nucleus of a gamete has the DNA content of 1C. A diploid nucleus from anaphase through G_1 has a 2C amount of DNA. This is duplicated during the S period so that from G_2 to metaphase the DNA content of the cell is 4C. It is more practical to talk about 2C and 4C nuclei than about diploid and tetraploid cells, since 4C represents both the G_2 phase of a diploid nucleus and the G_1 stage of a tetraploid one (Patau and Das, 1961).

Chromosome Replication

The mechanics of chromosome reproduction can be studied by autoradiographic techniques, which will be described in more detail in Chapter IV. Cells are grown for one S period in medium containing a radioactive constituent of DNA, usually ^{3}H-thymidine, and are then transferred back to normal medium. The results of such an experiment can be seen in Figs. III.2 and III.3 (cf. Taylor, 1963). Before the S period, the chromosome consists of one double helix of DNA. During the S period, each strand acts as a template for a new radioactive strand. In the subsequent metaphase each chromatid, which now has one "cold" and one "hot" DNA strand, is covered with silver grains. In the following S period, both hot and cold strands act as templates for new cold strands. In metaphase, one chromatid of each chromosome displays silver grains, the other does not. In *diplochromosomes*, which have undergone the same two syntheses without an intervening mitosis and chromatid segregation, the outer chromatids are radioactive (Fig. III.2). This shows that the new chromatids are the innermost in a diplochromosome. In the third diploid mitosis after the ^{3}H-thymidine treatment, the hot and the cold chromatids are distributed at random among the chromosomes (Fig. III.2). This type of chromosome reproduction is called *semiconservative*.

In Fig. III.3 semiconservative replication of human chromosomes cor-

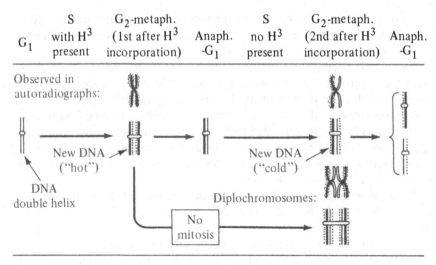

Conclusions: (i) at least one double helix per chromatid
(ii) linear organization at centromere: new DNA at outside

Figure III.2. Semiconservative replication of eukaryotic chromosomes demonstrated with tritiated thymidine and autoradiography.

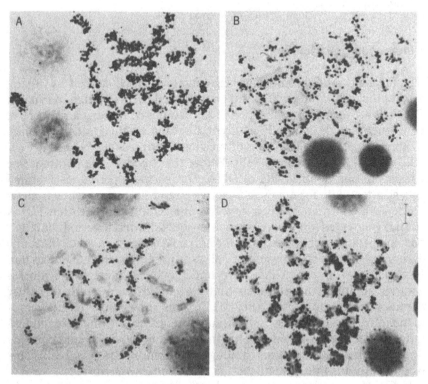

FIGURE III.3. Semiconservative replication of human chromosomes corresponding to Fig. III.2 (see text).

responding to Fig. III.2 is shown. Figure III.3.A illustrates a metaphase after one S period in "hot" medium. In Fig. III.3.B the chromosomes have been for one S period in "hot" medium and for another in "cold" medium. Figure III.3.C again illustrates a metaphase after one cycle in "hot" and two cycles in "cold" medium, and finally Fig. III.3.D shows diplochromosomes in which the outer chromatids are labelled.

Another characteristic feature in the replication of human and other eukaryotic chromosomes is that the DNA synthesis starts at several points along the chromosome. This can be demonstrated by giving the cells a short pulse of ^{3}H-thymidine, after which they are grown in cold medium. In metaphase the chromosome shows labeling over separate short stretches. The various chromosome segments start their replication at different times during the S period, with the heterochromatic segments of chromosomes being the last to replicate. The replication of chromosome segments can be studied more accurately by the more recent 5-bromodeoxyuridine techniques (cf. Latt, 1979).

References

Cremer T, Cremer C, Baumann H, et al (1982) Rabl's model of the interphase chromosome arrangement tested in Chinese hamster cells by premature chromosome condensation and laser-UV-microbeam experiments. Hum Genet 60: 46–56

Latt SA (1979) Patterns of late replication in human X chromosomes. In: Vallet HL, Porter IH (eds) Genetic mechanisms of sexual development. Academic, New York, pp 305–329

Patau K, Das NK (1961) The relation of DNA synthesis and mitosis in tobacco pith tissue cultured in vitro. Chromosoma 11: 553–572

Sperling K (1982) Cell cycle and chromosome cycle: Morphological and functional aspects. In: Rao PN, Johnson RT, Sperling K (eds) Premature chromosome condensation. Academic, New York, pp 43–78

Sperling K, Lüdtke E-K (1981) Arrangement of prematurely condensed chromosomes in cultured cells and lymphocytes of the Indian muntjac. Chromosoma 83: 541–553

Taylor JH (1963) The replication and organization of DNA in chromosomes. In: Taylor JH (ed) Molecular genetics I. Academic, New York, pp 65–111

Therman E, Denniston C (1984) Random arrangement of chromosomes in Uvularia (Liliaceae). Pl Syst Evol 147: 289–297

White MJD (1973) Animal cytology and evolution, 3rd edn. University Press, Cambridge, England

IV
Methods in Human Cytogenetics

In this chapter various available human chromosome techniques are reviewed. For the detailed descriptions the reader is referred to Darlington and La Cour (1976), to *Human Chromosome Methodology* (Yunis, 1965, 1974), Dutrillaux and Lejeune (1975), and to Yunis (1981b).

Direct Methods

In the 1920s and earlier, work on human chromosomes was done mostly on testicular tissue, on paraffin-sectioned preparations stained with hematoxylin. One can only admire the fact that with these relatively primitive techniques the human cytologists came as close to the right chromosome number as 48.

In the late 1940s the Feulgen squash technique came into general use (cf. Darlington and La Cour, 1976). It was an excellent method for studying unusual mitoses—for instance, mitotic aberrations in cancer cells. The chromosome constitution can also be analyzed in Feulgen squash preparations, if the number of chromosomes is not too large and the metaphase plate not too crowded. The root tips and pollen mother cells of many plants, as well as the testicular tissue of animals, have been studied successfully by this method.

The squash technique was greatly improved by the added step of treating the cells with drugs before fixation. These drugs, particularly colchicine, shorten the chromosomes, destroy the mitotic spindle, and as a result spread the chromosomes around the cell. Colchicine also prevents cells from entering anaphase; this leads to an accumulation of metaphases, a distinct advantage in cytological work. The most common fixative for Feulgen preparations, as well as for cells in tissue culture, is acetic acid-ethanol.

Tissue Culture Techniques

The real breakthrough in human cytogenetics came when tissue culture was combined with colchicine and hypotonic solution treatments before fixation. Thus, Tjio and Levan (1956) were able to report that in cultured embryonic lung cells the human chromosome number was 46. It should be pointed out, to the chagrin of everyone who had already been working on human chromosomes—including this author—that colchicine had been used to spread plant chromosomes and to produce polyploidy in plant breeding for two decades and that it had been known since the last century that hypotonic solutions swell animal cells.

Until 1960, when Moorhead et al launched the short-term lymphocyte culture technique, human chromosomes were studied mainly in cultured fibroblasts and in bone marrow cells. The essential agent in lymphocyte cultures is a phytohemagglutinin, the most widely used being an extract of kidney beans. Leukemic lymphocytes divide in vitro without "phyto", as it is usually called, whereas normal lymphocytes do not. Short-term lymphocyte cultures provide an important tool in human chromosome studies and have also found wide use in the cytogenetic studies of diverse animals. It takes only 48–72 h to obtain chromosome preparations with this easy method, whereas weeks of growth are needed before chromosomes can be studied in fibroblast cultures.

Drying of the cells on the microscope slide was a further improvement in that the flattened chromosomes are more easily stained and photographed. The most common chromosome stains have been orcein and Giemsa, but azur A and Feulgen as well as a few others have also been used.

At present, cultured lymphocytes and fibroblasts are the main sources of cells for human chromosome studies. However, both types of cultured cells have the disadvantage that they do not reveal what happens in vivo; this can be seen in mitoses from direct bone marrow biopsies, which however, usually do not yield chromosome preparations as beautiful as those from cultured cells.

Prenatal Studies

Amniocentesis, an important tool in genetic counseling, involves taking a sample of amniotic fluid, usually from a second trimester pregnancy. The sex of the embryo can be determined directly from such cells (which is important when counseling is done for diseases caused by X-linked genes), or tissue cultures for chromosome studies can be initiated from these cells. Certain metabolic diseases can also be determined by bio-chemical tests of fetal cells. If the analysis reveals a chromosome or bio-chemical anomaly that would lead to serious phenotypic consequences, an abortion can be performed. The various issues in the prenatal diag-nosis of genetic diseases have been reviewed by Epstein et al (1983).

A recent technique, still under development, that may finally replace amniocentesis is based on biopsies of trophoblastic villi. A catheter guided by ultrasound is inserted through the vagina into the uterus, and a villus sample is aspirated. Chromosomes (or enzymes) can be studied directly from such a biopsy or from a tissue culture grown from it. A great advantage of this method compared with amniocentesis is that it can be done during the first trimester and is less hazardous (cf. Simoni et al, 1983; Gustavii, 1983); moreover, abortions induced early in a pregnancy are less risky and more acceptable than those done later.

Meiotic Studies

Meiosis has been studied far more extensively in the human male than in the female. Initially male meiosis was analyzed from biopsies made into Feulgen squash preparations. A more recent method has been to pre-pare the meiocytes from the testicular tubules and to swell them in hypo-tonic solution, after which they are fixed and dried on a slide in the same way as lymphocytes.

With these techniques, elegant studies on the frequency and location of chiasmata in the human male (Fig. XV.8) have been made by Hultén and her colleagues (cf. Hultén, 1974; Hultén et al, 1978, 1982).

Synaptonemal complexes (Chapter XV) were originally studied solely by means of electron microscopy. However, microspreading of the chro-mosomes (cf. Solari, 1980), combined with silver staining (Pathak and Hsu, 1979) has also made their analysis possible in the light microscope (Fig. XV.4).

Female mammalian meiosis has been analyzed especially in the mouse. Similar attempts have been made in the human (Edwards, 1970; Jagiello et al, 1976), but so far the results have not been as satisfactory as those obtained in the mouse. Meiotic chromosome preparations of the mouse and other animals can be improved by an injection of colchicine into the animal a few hours before it is sacrificed. Meiotic prophase stages have been studied successfully in human embryonic oocytes (cf. Therman and Sarto, 1977). A special technique for the dictyotene (diplotene) stage in man has been developed by Stahl et al (1976).

The results of meiosis in the human male can now be analyzed by fer-tilizing hamster eggs, which have been freed from the zona pellucida, with human sperm and studying the haploid chromosome set in the male pro-nucleus (Rudak et al, 1978). This technique permits the determination of the type and frequency of chromosome aberrations in the sperm (for example, Martin et al, 1983; Brandriff et al, 1984).

Sex Chromatin Techniques

Since all the human cell types that are generally cultured are of mesodermal origin, practically the only information on the ectoderm and endoderm comes from studies on sex chromatin. This includes the X chromatin (Barr body), which consists of the inactive X, and the Y chromatin body, which represents the distal end of Yq. These are most often determined in cell smears made from the buccal mucosa. Barr bodies are usually stained with orcein, acid fuchsin, or Feulgen (Fig. XVIII.1), but the fluorochrome acridine orange also gives excellent results (cf. Fig. 2 in Dutrillaux, 1977). Because the Y body is too small to be distinct in the light microscope, it is usually stained with quinacrine and studied with fluorescence microscopy (Fig. XVII.1c). The fluorescent Y body is also visible in Y-carrying sperm. In addition to buccal smears, X chromatin has been analyzed in cultured fibroblasts, hair-root cells, and cells of the vaginal epithelium.

Autoradiography

As already discussed, only the chromosomes 1, 2, 3, 16, 17, 18, and the Y can be individually distinguished in the light microscope, the rest being identifiable only as members of a group. The realization that chromosomes and chromosome segments replicate at different times during the S period represented a considerable step forward in chromosome identification.

For autoradiography, the cells are fed a radioactive nucleotide (usually ^{3}H-thymidine) 5–9 h before fixation. This means that most cells observed in metaphase are then in the latter part of the S period. Those chromosome segments still replicating take up the radioactive thymidine. The cells are fixed and treated as usual, and a number of suitable metaphases are photographed. The slides are dipped in photographic emulsion, exposed in the dark, and developed. The previously photographed cells are rephotographed if they show a suitable number of silver grains (Fig. XVIII.2), which indicate the uptake of the radioactive compound. This method has been important for distinguishing the inactive X chromosome(s) from the active one. Some other chromosomes, for instance 21 and 22, can also be separated because 21 replicates later.

However, autoradiography, which is both time-consuming and inaccurate, has been largely replaced by banding techniques (for chromosome identification) and by the recent bromodeoxyuridine (BrdU) techniques (for the analysis of chromosome replication). Autoradiography is still used in cytogenetics to determine the timing of DNA synthesis and the duration of the mitotic stages.

Autoradiography plays an important role in the in situ hybridization techniques for the localization of specific, clustered, repeated DNA sequences on the chromosomes. This is done by hybridizing either ^{3}H-DNA labeled in vivo or ^{3}H-RNA transcribed in vitro with bacterial RNA polymerase directly on the chromosome slides in which the DNA strands have been separated by "denaturation" (Pardue and Gall, 1970). Thereafter autoradiography is performed in the usual way, and the location of the silver grains indicates the site of the specific DNA (Pardue, 1975). Single genes can also be localized with these techniques (Chapter XXVIII).

Banding Techniques of Fixed Chromosomes

A decisive step forward in human cytogenetics was the invention of banding techniques that differentiate the chromosomes into transverse bands of different lengths. With these methods, all the chromosomes of man and many other organisms, and even the breakpoints in most structural rearrangements, can be identified. The banding techniques and what they reveal about chromosome structure have been reviewed in many articles, for example, Arrighi (1974), Latt (1976), Evans (1977), Sanchez and Yunis (1977), Dutrillaux(1977), and Ris and Korenberg (1979).

When human chromosomes are stained with quinacrine HCl (Atebrine) or quinacrine mustard and studied with a fluorescence microscope, they show bands of different brightness (Caspersson et al, 1970). These are called Q-bands (quinacrine bands) (Fig. XXI.3). Slides stained with quinacrine are not permanent, and after a couple of photomicrographic exposures the fluorescence fades too much to be usable.

The other type of banding technique involves various pretreatments and staining with Giemsa or certain fluorochromes; Giemsa bands (G-bands) are obtained when the chromosomes are pretreated with a salt solution at 60°C or with proteolytic enzymes, usually trypsin (Drets and Shaw, 1971). Giemsa banding yields essentially the same information as Q-banding, only the brightly fluorescent Q-bands are now darkly stained, whereas the Q dark regions are light (Fig. II.4). Each method has its advantages. With Q-banding the chromosomes are stained without any pretreatment, and their morphology is retained. This makes measurements of bands more accurate. Also the relative brightness of the bands can be estimated (Kuhn, 1976). However, the chromosomes can be analyzed only from photographs.

The G-banded slides, on the other hand, are permanent and are therefore more suitable for routine work. By means of these two techniques, some 300 bands have been described in the human chromosomes (Figs. IV.1 and 2) (Paris Conference, 1971). However, in most metaphase plates only a fraction of this number of bands can actually be seen. By accumulation of cells in prophase and by banding these long chromosomes, the number of visible bands has been increased to 1000–2000 (see below).

Reverse banding (R-banding) involves pretreatment with hot (80–90°C) alkali and subsequent staining with Giemsa or a fluorochrome. As the name indicates, the banding pattern is the reverse of G-banding; in other words, the bands that are dark with R-banding are light with G-banding and vice versa (Fig. V.1f). Fluorescent R-banding again is the reverse of Q-banding. Although this banding reveals nothing new compared with Q-banding and G-banding, it complements them when chromosome ends are especially studied, as in distal deletions and translocations. A modification of R-banding, called T-banding, brings out mainly the tips of chromosomes.

Another technique, which also utilizes the Giemsa stain, has given additional information about chromosome structure. Centric banding (C-banding), for which chromosomes are usually first treated in acid and then in alkali (barium hydroxide, for example) prior to Giemsa staining, brings out the heterochromatic regions around the centromere (Fig. V.1f and g) and in the distal end of the Y chromosome (Arrighi and Hsu, 1971). With a modification of the C-banding technique (G–11), it is possible to stain the centric heterochromatin in human chromosome 9 specifically.

The banding technique used most successfully in plants corresponds to C-banding, and the resulting darkly stained bands probably also represent constitutive heterochromatin (Chapter V).

New banding techniques, often combining two or more stains, appear frequently. A variety of fluorochromes, including Hoechst 33258, a number of antibiotics, and so-called DAPI and DIPI stains, have found use in the detailed analysis of chromosome bands (cf. Schweizer, 1981).

With a special Giemsa method it is possible to stain nucleolar organizers differentially, and a silver (NOR) staining technique reveals the organizers that are active (Fig. V.1c and e) (cf. Bloom and Goodpasture, 1976).

So-called Cd banding stains active centromeres, but not inactive ones (Eiberg, 1974).

Banding Techniques for Cells in Culture

The banding techniques that involve the treatment of cells before fixation are based on the important discovery that chromosomes and chromatids that have incorporated BrdU have a different structure and consequently different staining properties from those containing thymidine. One of the applications of this observation is to cause one chromatid to incorporate BrdU, whereas the other contains thymidine. Sister chromatid exchanges, the existence of which was already revealed by autoradiography, can be studied with much greater accuracy with this technique (Latt, 1973). For the analysis of sister chromatid exchanges, the fluorochromes Hoechst 33258, acridine orange, and coriphosphine O, as well as the Giemsa stain after heat treatment, have been used (Korenberg and Freedlender, 1974).

Substitution with BrdU has also been successful in demonstrating the late-labeling X chromosome and other individual chromosomes and segments that replicate in the latter part of the S period (Latt, 1973, 1974). This can be done, for instance, by growing lymphocytes for 40–44 h in a medium containing BrdU and, thereafter, feeding them thymidine for 6–7 h. When stained with a suitable fluorochrome, the late-replicating X chromosome fluoresces more brightly than the other chromosomes or is darker than the others when Giemsa staining is used. An example of this technique is seen in Fig. XVIII.3, in which the late-replicating X chromosomes stand out.

This technique can also be done in reverse in that the cells are first grown in thymidine-containing medium and then with BrdU. The late-replicating chromosomes, including the inactive X chromosome, are now darkly fluorescent with quinacrine and faintly stained with Giemsa.

Prophase Banding

One of the most important recent developments in cytogenetic techniques is the prophase or high-resolution banding (Yunis, 1976). Whereas the usual Q-, G-, or R-banding techniques reveal some 300 bands, prometaphase or prophase chromosomes show from 500 to 2000 bands (cf. Yunis et al, 1979; Yunis, 1981a). For high-resolution banding, dividing cells are blocked at the S stage with amethopterin (methotrexate). When the block is released with thymidine-rich medium, a large proportion of the cells are synchronous and can be fixed at the required stage. Prophase chromosomes show natural banding with Giemsa or Wright's stain, or the banding can be enhanced with G-banding techniques.

The longer the chromosomes, the more bands they show. However, the more bands, the more overlapping and the more tedious the analysis becomes. The most important use of prophase banding has been the exact determination of breakpoints in cases in which the general area of the breaks is already known.

Nomenclature of Human Chromosomes

As a continuation of previous agreements on the nomenclature of human chromosomes, at the Paris Conference (1971) a system was proposed for identifying human chromosome bands and indicating various chromosome abnormalities. Figure IV.1 shows a diagram of a banded human

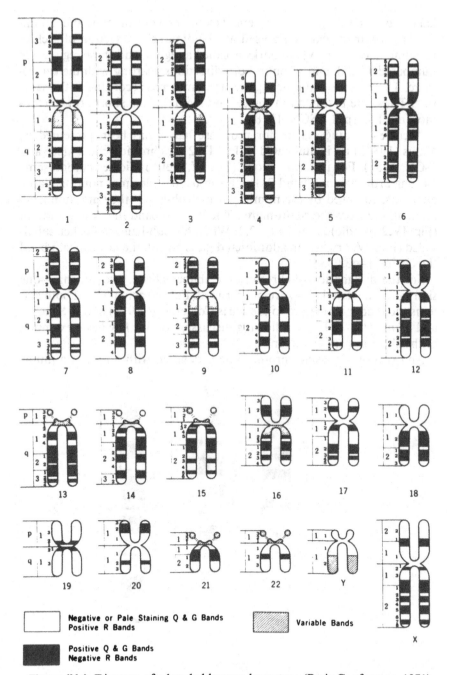

Figure IV.1. Diagram of a banded human karyotype (Paris Conference, 1971).

karyotype according to this system. Telomeres, centromeres, and a number of prominent bands are used as "landmarks". A section of a chromosome between two landmarks is called a region, and these regions are numbered 1, 2, 3, and so on in both directions, starting with the centromere. The bands within the regions are numbered according to the same rule. Thus, the first band in the second region of the short arm of chromosome 1 is 1p21 (ISCN, 1978).

A corresponding system for prophase bands has been presented by Yunis (1980) and is illustrated in Fig. IV.2 for chromosome 14 (see also ISCN, 1981). For instance, 14q32 (Fig. IV.2 left) indicates chromosome 14 long arm region 3, band 2. For a sub-band, a decimal point is placed after this, followed by the number of the sub-band (they are numbered sequentially from the centromere). The last sub-band in chromosome 14 (Fig. IV.2 middle) is thus 14q32.3. When the sub-band is further subdivided (Fig. IV.2 right), an additional digit is added, the last sub-sub-band thus being 14q32.33.

For the designation of chromosome abnormalities, two systems—one short and one detailed—were put forward. For the actual use of both systems the reader is referred to the Paris Conference (1971) and to Sanchez and Yunis (1977). In the following discussion only a few basic examples of the short system are given.

An extra or a missing chromosome is denoted with a plus or a minus

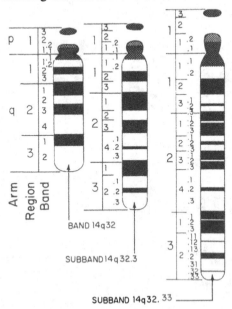

Figure IV.2. Schematic representation of chromosome #14 at the 320 (left), 500 (center), and 900 band stage (right), illustrating the use of the Paris nomenclature to designate with decimals the sub-bands seen in the 500 band stage, and with decimals and digits sub-bands observed in the 900 band stage (Yunis, 1980).

sign, respectively, before the number of the chromosome. Thus the chromosome constitution of a female with trisomy for chromosome 13 would be 47,XX,+13, and a male with monosomy for 21 would have the formula 45,XY,−21.

A plus or minus sign after the symbol of the chromosome arm means that a segment is added to or missing from it. For example, a female with the cri du chat syndrome would have the chromosome formula 46,XX,5p−, and a male with an abnormally long 4q would be designated as 46,XY,4q+. The karyotype of a female with a Robertsonian translocation (centric fusion) between chromosomes 13 and 14 would be 45,XX,t(13q14q). The formula for a male carrier in whom chromosome arms 3p and 6q have exchanged segments, the breakpoints being 3p12 and 6q34, would be 46,XY,t(3;6)(p12;q34).

Unfortunately the lengths of chromosome bands in Fig. IV.1 (Paris Conference, 1971) are not based on actual measurements, although a realistic banded diagram would be most useful. Furthermore, only a small fraction of metaphase plates reveal all the bands illustrated in this diagram. Therefore it is often difficult, if not impossible, to determine the exact breakpoints in, say, a reciprocal translocation. Despite this, most authors feel obliged to specify the breakpoints, even when they are based only on guesses. Obviously, it will be possible to determine breakpoints more accurately in the future, when the banding of prophase chromosomes comes into wider use (cf. Sanchez and Yunis, 1977).

Quantitative Methods

One of the most important applications of quantitative methods in cytogenetics has been the measuring of the DNA content of nondividing nuclei; for instance, in differentiated or malignant cells or in sperm. In spectrocytophotometry, which has been used for over 30 years, the light absorption of nuclei or chromosomes is determined (Patau, 1952). This has been done most successfully on stained nuclei, and the stain most often used has been Feulgen (cf. Mendelsohn, 1966). The light absorption of individual chromosomes or chromosome bands can be done in the same way (cf. Drets and Seuánez, 1973).

A more recent method, which for many purposes has replaced cytophotometry, is flow cytometry, the cytogenetic applications of which seem to be almost limitless. Its essential feature is that suspended single cells (or chromosomes) pass a detector by means of a flow channel (cf. Mendelsohn, 1980). For the determination of the DNA content of nuclei and chromosomes, flow cytometry is a much faster technique than cytophotometry. Another function of flow cytometry is the sorting out of similar-sized nuclei or chromosomes; this technique has found important applications, for instance, in gene mapping (cf. Yu et al, 1984).

References

Arrighi FE (1974) Mammalian chromosomes. In: Busch H (ed) The cell nucleus. Academic, New York, pp 1–32

Arrighi FE, Hsu TC (1971) Localization of heterochromatin in human chromosomes. Cytogenetics 10: 81–86

Bloom SE, Goodpasture C (1976) An improved technique for selective silver staining of nucleolar organizer regions in human chromosomes. Hum Genet 34: 199–206

Brandriff B, Gordon L, Ashworth L, et al (1984) Chromosomal abnormalities in human sperm: Comparisons among four healthy men. Hum Genet 66: 193–201

Caspersson T, Zech L, Johansson C (1970) Differential binding of alkylating fluorochromes in human chromosomes. Exp Cell Res 60: 315–319

Darlington CD, La Cour LF (1976) The handling of chromosomes, 6th edn. Wiley, New York

Drets ME, Seuánez H (1973) Quantitation of heterogeneous human heterochromatin: Microdensitometric analysis of C- and G-bands. In: Coutinho EM, Fuchs F (eds) Physiology and genetics of reproduction, Part A. Plenum, New York, pp 29–52

Drets ME, Shaw MW (1971) Specific banding patterns of human chromosomes. Proc Natl Acad Sci USA 68: 2073–2077

Dutrillaux B (1977) New chromosome techniques. In: Yunis JJ (ed) Molecular structure of human chromosomes. Academic, New York, pp 233–265

Dutrillaux B, Lejeune J (1975) New techniques in the study of human chromosomes: Methods and applications. In: Harris H, Hirschhorn K (eds) Advances in human genetics, Vol. 5. Plenum, New York, pp 119–156

Edwards RG (1970) Observations on meiosis in normal males and females. In: Jacobs PA, Price WH, Law P (eds) Human population cytogenetics. Williams and Wilkins, Baltimore, pp 10–21

Eiberg H (1974) New selective Giemsa technique for human chromosomes, Cd staining. Nature 248: 55

Epstein CJ, Cox DR, Schonberg SA, et al (1983) Recent developments in the prenatal diagnosis of genetic diseases and birth defects. Annu Rev Genet 17: 49–83

Evans HJ (1977) Some facts and fancies relating to chromosome structure in man. In: Harris H, Hirschhorn K (eds) Advances in human genetics, Vol. 8. Plenum, New York, pp 347–438

Gustavii B (1983) First-trimester chromosomal analysis of chorionic villi obtained by direct vision technique. Lancet ii: 507–508

Hultén M (1974) Chiasma distribution at diakinesis in the normal human male. Hereditas 76: 55–78

Hultén M, Luciani JM, Kirton V, et al (1978) The use and limitations of chiasma scoring with reference to human genetic mapping. Cytogenet Cell Genet 22: 37–58

Hultén MA, Palmer RW, Laurie DA (1982) Chiasma derived genetic maps and recombination fractions: Chromosome 1. Ann Hum Genet 46: 167–175

ISCN (1978) An international system for human cytogenetic nomenclature. Cytogenet Cell Genet 21: 309–404

ISCN (1981) An international system for human cytogenetic nomenclature—high resolution banding. Cytogenet Cell Genet 31: 1–84

Jagiello G, Ducayen M, Fang J-S, et al (1976) Cytogenetic observations in mammalian oocytes. In: Pearson PL, Lewis KR (eds) Chromosomes today, Vol. 5. Wiley, New York, pp 43–63

Korenberg JR, Freedlender EF (1974) Giemsa technique for the detection of sister chromatid exchanges. Chromosoma 48: 355–360

Kuhn EM (1976) Localization by Q-banding of mitotic chiasmata in cases of Bloom's syndrome. Chromosoma 57: 1–11

Latt SA (1973) Microfluorometric detection of deoxyribonucleic acid replication in human metaphase chromosomes. Proc Natl Acad Sci USA 70: 3395–3399

Latt SA (1974) Microfluorometric analysis of DNA replication in human X chromosomes. Exp Cell Res 86: 412–415

Latt SA (1976) Optical studies of metaphase chromosome organization. Annu Rev Biophys Bioeng 5: 1–37

Martin RH, Balkan W, Burns K, et al (1983) The chromosome constitution of 1000 spermatozoa. Hum Genet 63: 305–309

Mendelsohn ML (1966) Absorption cytophotometry. Comparative methodology for heterogeneous objects and the two-wavelength method. In: Wied GL (ed) Introduction to quantitative cytochemistry. Academic, New York, pp 201–237

Mendelsohn ML (1980) The attributes and applications of flow cytometry. Flow Cytom 4: 15–27

Moorhead PS, Nowell PC, Mellman WJ, et al (1960) Chromosome preparations of leukocytes cultured from human peripheral blood. Exp Cell Res 20: 613–616

Pardue ML (1975) Repeated DNA sequences in the chromosomes of higher organisms. Genetics 79: 159–170

Pardue ML, Gall JG (1970) Chromosomal localization of mouse satellite DNA. Science 168: 1356–1358

Paris Conference (1971) Standardization in human cytogenetics. Birth Defects: original Article Series, VIII: 7. The National Foundation, New York, 1972

Patau K (1952) Absorption microphotometry of irregular shaped objects. Chromosoma 5: 341–362

Pathak S, Hsu TC (1979) Silver-stained structures in mammalian meiotic prophase. Chromosoma 70: 195–203

Ris H, Korenberg JR (1979) Chromosome structure and levels of chromosome organization. In: Goldstein L, Prescott DM (eds) Cell biology, Vol. 2. Academic, New York, pp 268–361

Rudak E, Jacobs PA, Yanagimachi R (1978) Direct analysis of the chromosome constitution of human spermatozoa. Nature 274: 911–913

Sanchez O, Yunis JJ (1977) New chromosome techniques and their medical applications. In: Yunis JJ (ed) New chromosomal syndromes. Academic, New York, pp 1–54

Schweizer D (1981) Counterstain-enhanced chromosome banding. Hum Genet 57: 1–14

Simoni G, Brambati B, Danesino C, et al (1983) Efficient direct chromosome analyses and enzyme determinations from chorionic villi samples in the first trimester of pregnancy. Hum Genet 63: 349–357

Solari AJ (1980) Synaptonemal complexes and associated structures in microspread human spermatocytes. Chromosoma 81: 315–337

Stahl A, Luciani JM, Gagné R, et al (1976) Heterochromatin, micronucleoli, and RNA containing body in the pachytene and diplotene stages of the human oocyte. In: Pearson PL, Lewis KR (eds) Chromosomes today, Vol. 5. Wiley, New York, pp 65–73

Therman E, Sarto GE (1977) Premeiotic and early meiotic stages in the pollen mother cells of *Eremurus* and in human embryonic oocytes. Hum Genet 35: 137–151

Tjio JH, Levan A (1956) The chromosome number of man. Hereditas 42: 1–6

Yu L-C, Gray JW, Langlois R, et al (1984) Human chromosome karyotyping and molecular biology by flow cytometry. In: Sparkes RS, de la Cruz FF (eds) Research perspectives in cytogenetics. University Park, Baltimore, pp 65–73

Yunis JJ (ed) (1965, 1974) Human chromosome methodology, 1st and 2nd edns. Academic, New York

Yunis JJ (1976) High resolution of human chromosomes. Science 191: 1268–1270

Yunis JJ (1980) Nomenclature for high resolution human chromosomes. Cancer Genet Cytogenet 2: 221–229

Yunis JJ (1981a) Mid-prophase human chromosomes.The attainment of 2000 bands. Hum Genet 56: 293–298

Yunis JJ (1981b) New chromosome techniques in the study of human neoplasia. Hum Pathol 12: 540–549

Yunis JJ, Ball DW, Sawyer JR (1979) G-banding patterns of high-resolution human chromosomes 6–22, X, and Y. Hum Genet 49: 291–306

V
Longitudinal Differentiation of Eukaryotic Chromosomes

Longitudinal Differentiation of Chromosomes

Longitudinal differentiation of chromosomes is expressed at several different levels. However, there are still gaps in the information gained by different methods, and the goal of future studies will be to close these gaps. Observations on nonbanded chromosomes have been replaced to a large extent by studies on variously banded chromosomes. Molecular studies have shown that chromosomes consist of repetitive, middle repetitive, and unique DNA sequences; a picture of the fine structure of the eukaryotic chromosome is finally emerging (Chapter VI).

Molecular Differentiation of Chromosomes

On the molecular level the genome of eukaryotes consists of highly repeated sequences ($>10^5$ copies), middle repeated sequences (ca. 10^2–10^4 copies) and unique sequences (one or a few copies) (cf. Hood et al, 1975).

In the highly repeated sequences, which include constitutive heterochromatin, the sequence complexities vary from 2 to at least 2350 base pairs (cf. Miklos, 1982), and the sequences vary extensively both within and between species. So far no definite biological role has been found either for constitutive heterochromatin, which is never transcribed, or for the variations that occur in it.

Multigene families, which code, for instance, for ribosomal RNA or histones, represent middle repeated DNA. The Alu-I family, consisting of 300 base pairs, is repeated some 300,000 times dispersed throughout the human genome. Furthermore, transposable elements, which do not have a fixed location on the chromosomes, make up a considerable proportion of the middle repeated DNA in such organisms as yeast and *Drosophila* (cf. Finnegan et al, 1982), and this may also be true elsewhere.

Unique sequences can be divided into transcribed genes and noncoding unique sequences, to which spacers and so-called pseudogenes also belong.

The informational multigene families, of which globin and immunoglobulin genes are typical examples, bridge the gap between unique sequences and middle repeated sequences. Characteristic of genes in such multigene families is multiplicity, close linkage, sequence homology, and similar or overlapping functions (cf. Hood et al, 1975).

Prebanding Studies

Apart from primary and secondary constrictions and satellites, chromosomes of most higher organisms display differential segments, sometimes whole chromosomes, that exhibit a condensation cycle which deviates from the main part of the chromosomes. Heitz (for example, 1933), who made a number of basic studies in this field during the late 1920s and early 1930s, named these chromosome parts *heterochromatic* in contrast to the majority of the chromosomes, which he termed *euchromatic* (cf. Passarge, 1979). Chromosomes that are more condensed than the rest have also been called *positively heteropycnotic*, whereas those that are less condensed are *negatively heteropycnotic*. The same chromosome may at some stages be positively, at others negatively, heteropycnotic. The X chromosome in the spermatocytes of grasshoppers is an example of this.

Being out of step with the rest of the chromosomes, which is one of the characteristics of heterochromatin, is called *allocycly*. In interphase, heterochromatic chromosome segments often appear as more condensed *chromocenters* or *prochromosomes* (cf. Wilson, 1925). As a result of the studies of Heitz, it is known that heterochromatin is genetically inert. The prebanding knowledge of heterochromatin has been reviewed by Vanderlyn (1949).

The first successful experiments designed to make heterochromatic segments visible in metaphase chromosomes were completed in the late 1930s when Darlington and La Cour (1940) discovered differential chromosome regions in the liliaceous plants *Paris* and *Trillium*; these regions, after cold treatment, appeared thinner and less stained than the rest of the chromosomes. In interphase nuclei the same segments were visible as positively heteropycnotic chromocenters. This phenomenon has been observed in several plants and a few animal species. Darlington and La Cour (1940) pointed out that these differential segments had to be genetically inert, since they varied from plant to plant, and many individuals were heterozygous for them.

A forerunner of the present banding techniques was the acid treatment combined with orcein staining of Yamasaki (1956); this produced excellent banding in orchid chromosomes. However, Yamasaki's interesting work appeared before the time was ripe for it, and it did not attract the attention it deserved.

Banding Studies on Human Chromosomes

Q-banding differentiates the human chromosomes into bands of differing length and relative brightness. These two parameters were determined by Kuhn (1976). An interesting property of the metaphasic Q-bands is that they seem to act as units; for instance, each band replicates within a particular limited time during the S period (cf. Ris and Korenberg, 1979). Interestingly, the Q- and G-bands correspond to the chromomeres, as seen in pachytene. These bands have also shown amazing constancy during evolution. Because the overwhelming majority of even prophase bands are identical in both man and the great apes, one can conclude that they have remained unchanged for at least 20 million years, some of them for considerably longer (cf. Yunis and Prakash, 1982).

The most brightly fluorescent bands in the human chromosome complement are the distal end of the Y chromosome and the narrow variable bands at the centromeres of chromosomes 3, 4, and the acrocentrics. Human populations are polymorphic for these bright bands, just as they are for the size and fluorescent properties of the satellites.

G-banding and its reverse, R-banding, give essentially the same information as Q-banding. However, R-banding is often useful for the study of structural changes involving chromosome ends that might go undetected with Q-banding or G-banding.

C-banding reveals a type of chromatin that is in principle different from euchromatin. This technique specifically stains constitutive heterochromatin, which is situated at the centric regions of all human chromosomes and at the distal end of the Y chromosome. Particularly prominent blocks of this heterochromatin are found in chromosomes 1, 9, and 16, and these regions show considerable polymorphism in the population. Possibly a variation in the centric heterochromatin of these chromosomes is more noticeable than variation in the shorter C-bands; however, variation in some of the shorter bands has also been reported. The constitutive heterochromatin in chromosomes 1, 9, 16, and the Y varies from shorter than average to about three times its usual length.

Constitutive heterochromatin seems especially prone to breakage and rearrangement. However, for the most part the polymorphism of the C-bands is probably the result of unequal pairing and crossing-over. Nevertheless, these changes cannot take place very often, since the C-band variants as a rule are constant from one generation to the next and show normal mendelian inheritance.

Nucleoli and Chromocenters

The nucleolar organizers, which consist of ribosomal RNA genes, of which the human has about 200, are situated at the satellite stalks of the D and G chromosomes (cf. Miller, 1981). In many organisms these genes are localized in secondary constrictions of one (or more) chromosome pair as in *Haplopappus* (Fig. II.2a). Both satellite stalks and secondary constrictions are usually visible without any banding techniques. That

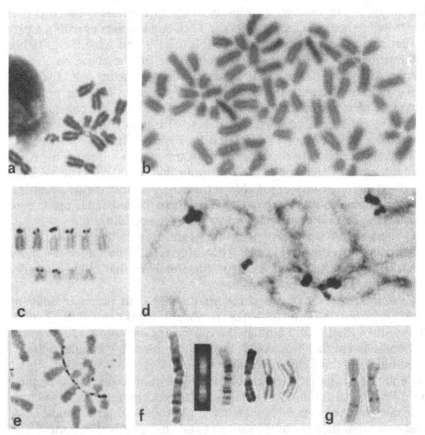

Figure V.1. (a) Satellite association of human acrocentric chromosomes; (b) centric association of mouse cancer chromosomes; (c) silver- staining of active nucleolar organizers in human D and G chromosomes (courtesy of C Trunca); (d) heterochromatic association of mouse diplotene bivalents; (e) satellite association of silver-stained human acrocentrics (courtesy of TM Schroeder); (f) human chromosome 1 (from the left): prometaphase banding, Q-banding, G-banding, R-banding, C-banding (from a heterozygote for the C-band); (g) C-banded human dicentric chromosomes with one centromere inactivated.

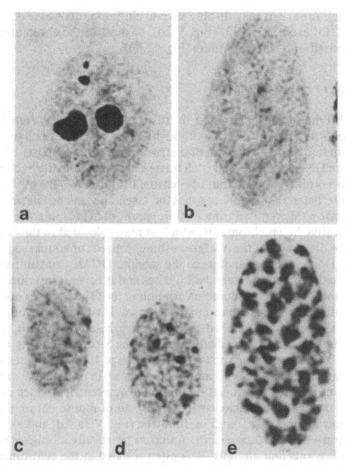

Figure V.2. Human interphase nuclei. (a) Nucleus from placenta; NOR-silver staining shows the nucleoli black; (b-e) Feulgen-stained nuclei from cervical cancer: (b) fairly evenly stained nucleus; (c) nucleus showing an X chromatin body; (d) nucleus with several chromocenters; (e) most of the chromatin condensed into chromocenters.

they contain the nucleolar organizers was discovered by Heitz (1931) and verified with in-situ hybridization techniques (Henderson et al, 1972). Active nucleolar organizers can be stained with the NOR silver technique (Fig. V.1c and e) (Goodpasture and Bloom, 1975), which also shows the nucleoli in interphase nuclei (Fig. V.2a). Each nucleolar organizer forms a nucleolus in telophase; the nucleoli fuse during interphase to disappear in prophase.

With the Feulgen technique, which does not stain the nucleoli, interphase nuclei may stain fairly evenly (Fig. V.2b), or a Barr body is visible

(Fig. V.2c). A nucleus may display several chromocenters, also described by Heitz (for example, 1933) (Fig. V.2d), or most of the chromatin may be condensed into chromocenters (Fig. V.2e).

Constitutive Heterochromatin

Heterochromatin can be divided into two main categories, *constitutive* and *facultative* (Brown, 1966). Although these two classes behave similarly in many ways, a fundamental structural difference exists between them. The DNA structure that is characteristic of the constitutive heterochromatin is different from that of euchromatic DNA. On the other hand, facultative heterochromatin consists of essentially euchromatic DNA, but it *behaves* differently during certain phases of development.

Constitutive heterochromatin, which in man is located in the C-bands and at the distal end of the Y chromosome, consists of simple sequence-repeated DNA that corresponds to the satellite DNAs. Constitutive heterochromatin contains no mendelian genes and is never transcribed. This explains the fact that considerable variation in the C-bands does not seem to affect the phenotype, even in extreme cases. Characteristics of constitutive heterochromatin, shared by facultative heterochromatin, are its genetic inertness, its late replication during the S period, and its general allocycly. It has long been known (cf. Vanderlyn, 1949) that heterochromatic segments are "sticky" and tend to fuse in interphase. This is reflected in the satellite association of human acrocentrics (which may be helped by remnants of nucleolar material) and of mouse chromosomes, both in mitotic metaphase and in diplotene (Fig. V.1a,b,d, and e).

Constitutive heterochromatin is found in practically all higher organisms, both plants and animals. It is often situated at the centromere, as in the human and the mouse. In many plants, for instance the onion, it forms blocks at the telomeres, but it can also be intercalary (Fig. V.3a). Before the invention of banding techniques, the deer mouse *Peromyscus* presented a cytological riddle. The length of its chromosomes and the centromere index seemed to vary from animal to animal, and complicated inversion systems were invented to explain this observation. C-banding revealed that all the short arms consisted of constitutive heterochromatin, the amount varying from animal to animal (cf. Hsu, 1975). Closely related species with the same chromosome number sometimes have very different amounts of DNA per nucleus. This, too, has been found to depend on variation in constitutive heterochromatin (cf. Gall, 1981).

The reaction of the different C-bands to the various banding techniques shows that the structure of their DNA is not identical. Thus, although all the constitutive heterochromatin stains dark with C-banding, the distal end of the Y chromosome is brightly fluorescent with Q-banding, whereas

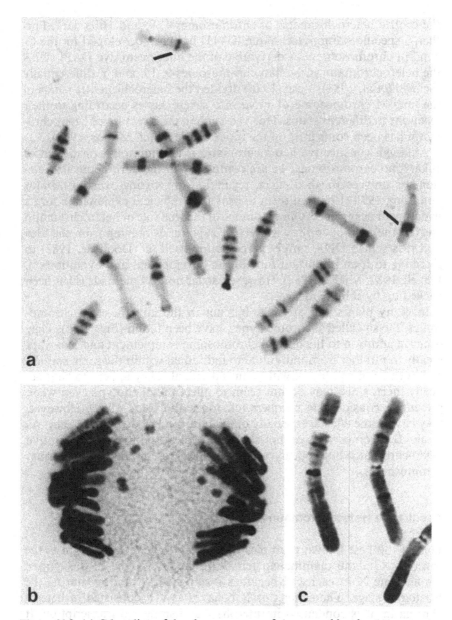

Figure V.3. (a) C-banding of the chromosomes of *Anemone blanda*; chromosome with heterozygous C-band marked (courtesy of D Schweizer); (b) B-chromosomes segregating irregularly in I meiotic anaphase of *Fritillaria imperialis* (courtesy of M Ulber); (c) G-banded chromosomes of *Pinus resinosa*; the chromosome on the left has a secondary constriction (courtesy of A Drewry).

the centric heterochromatin of chromosomes 1, 9, and 16 is dark. Further, a specific staining technique (G–11) has been developed for the C-band in chromosome 9. A derivative of the fluorescent dye DAPI stains the heterochromatin in human chromosomes 9, 15, and Y differentially (Schnedl et al, 1981). Vosa (1976) divided the heterochromatic bands in the long M chromosome of *Vicia* into seven classes according to their reactions to different stains. This variation in the structure of heterochromatin has been confirmed on the molecular level (cf. Miklos, 1982).

Although constitutive heterochromatin is a ubiquitous constituent of eukaryotic chromosomes, we are completely in the dark about its function. Its only established effect, regulation of crossing-over (cf. Miklos and John, 1979), does not seem to provide a sufficient explanation for its widespread occurrence. Our ignorance of the true role of heterochromatin has left the field open for a variety of hypotheses ranging from the idea that it is selfish DNA simply perpetuating itself (cf. Doolittle, 1982) to ascribing to it an important function in development and evolution (cf. Flavell, 1982; Miklos, 1982). However, so far none of these ideas has been backed up by solid evidence.

In many plants and animals, but not in the human, extra chromosomes, the so-called B chromosomes, have been found (Fig. V.3b). They occur in addition to the normal chromosome complement and may vary greatly in number from individual to individual within the same species, and even from cell to cell within an individual. They appear to be practically inert, since they do not seem to affect the phenotype even when present in considerable numbers (cf. Jones and Rees, 1982). However, they have some effects, especially on viability and chiasma frequency. As a rule, B chromosomes do not show the typical staining of constitutive heterochromatin but appear variably banded, resembling small, normal chromosomes.

Facultative Heterochromatin

One of the best known examples of facultative heterochromatin is the inactive X in the mammalian female (Chapter XVIII). In the human female, one X in each cell becomes inactivated at random during the blastocyst stage. Thereafter this X behaves as if it consisted of heterochromatin. It is condensed in interphase and shows no transcription. It replicates late during the S period (Figs. XVIII.2 and 3) and is more condensed than the other chromosomes, even in prophase and prometaphase (Fig. XVIII.2). Before inactivation, both X chromosomes are active, and the inactive X is turned on again before oogenesis, when both X chromosomes behave much like the autosomes (Chapter XVIII). During spermatogenesis, on the other hand, neither the X nor the Y chromosome is transcribed (Chapter XVIII). Both chromosomes behave like heterochromatin. The mechanism that determines the behavior of the facultative heterochromatin is still unknown.

Intercalary Heterochromatin

A third type of heterochromatin has been called *intercalary heterochromatin* (Patau, 1973). However, we know even less about it than about the other categories of heterochromatin. It is situated in the Q-bright chromosome bands, which are late replicating and seem to contain fewer genes than the dark bands.

Only a small part of the DNA present is transcribed in any cell (cf. Ris and Korenberg, 1979). No condensed parts, whether consisting of constitutive or facultative heterochromatin, code for any proteins. In some cell types, such as erythrocytes or sperm, in which the total chromatin is in a condensed state (cf. Ris and Korenberg, 1979), this inactivity applies to the whole chromosome complement. A different type of heterochromatic behavior is exhibited by chromosomes or chromosome segments that are condensed in differentiated cells, forming chromocenters of various shapes and sizes (Fig. V.2d and e). This is obviously a mechanism to shut off those genes whose activity is not needed at a certain point of development (cf. Ris and Korenberg, 1979). Plants and animals seem to differ in that in animals the condensation of chromatin in interphase acts as a regulating mechanism for gene action, whereas in plants the condensed parts represent constitutive heterochromatin (cf. Nagl, 1982). However, not only is *condensed* chromatin silent, but a great part of the *extended* chromatin is never transcribed in any cell type (Ris and Korenberg, 1979).

Chromosome Bands

As new banding techniques continually emerge, often combining different stains, a more detailed longitudinal differentiation of chromosomes becomes possible. The ongoing classification of chromosome bands has told us little about their possible functions or about the chemical basis of the differential staining reactions. The most information exists about Q-banding. Weisblum and de Haseth (1972) and Weisblum (1973) showed in their in vitro studies on polynucleotides of known composition that DNA consisting of repeated adenine-thymine (AT) base pairs fluoresces brightly with quinacrine, whereas guanine-cytosine (GC) pairs quench fluorescence. Quenching is also a function of the degree of the interspersion of GC pairs (AT pairs must occur in uninterrupted stretches of a certain length to cause fluorescence). Two stretches of DNA can have the same base ratio (AT:GC) but fluoresce differently, depending on the sequence of the bases.

Korenberg and Engels (1978), who were the first to determine directly the base ratios of the Q-bands in human chromosomes, demonstrated that the Q-brightness of the bands was positively correlated with the AT:GC ratio. However, the differences in the base ratios between the very bright Y heterochromatin and the dark bands are much less than has often been assumed (cf. Evans, 1977). Interestingly, the longer chromosomes are relatively brighter than the smaller ones. Although the DNA structure is the basis for Q-banding, the chromosomal proteins, both histones and nonhistones, obviously play an important role, especially in the various Giemsa techniques (cf. Ris and Korenberg, 1979). Nevertheless, our understanding of the chemical basis of the various banding techniques is just beginning.

Function of Human Chromosome Bands

One of the most intriguing questions concerning the chromosome bands is: What is the functional difference between Q-bright (G-dark) and Q-dark (G-light) bands?

Yunis (1965) pointed out that the only three autosomes (13, 18, and 21) for which trisomy is viable to any appreciable extent are the latest replicating chromosomes in their groups. Now we know that they are also the Q-brightest. Yunis concluded that these chromosomes contained more heterochromatin than did early replicating chromosomes; therefore, their trisomies were better tolerated than trisomies for more gene-rich autosomes. A similar nonrandomness has been shown for spontaneous trisomic abortions (Boué et al, 1976).

A variety of other observations support the idea that genes are not distributed at random along the chromosomes, but are concentrated in the Q-dark regions, some of which are special hot spots. Kuhn (1976) demonstrated that mitotic chiasmata in Bloom's syndrome (Chapter X) are localized extremely nonrandomly, Q-dark regions in 1p, 3p, 6p, 11q, 12q, 17q, 19 (p or q), and 22q being especially chiasma-rich. A strong negative correlation was also found between the hot spot chromosomes and their involvement in trisomic abortions (Korenberg et al, 1978; Kuhn et al, 1985). In other words, embryos that are trisomic for these chromosomes die too early to be recognized as abortions. The same chromosomes also have a significantly higher number of localized genes than similar-sized control chromosomes (Kuhn et al, 1985). Korenberg et al (1978) formulated the hypothesis that the Q-darker chromosome bands have higher gene densities than do the brighter ones, and the chiasma hot spots are especially prominent in this respect. Because they contain active genes, they would be extended in interphase and as a result would be more easily available for mitotic pairing and crossing-over.

Many other observations also speak for the relative gene-richness of the

Q-dark regions. For instance,trisomy for the short Q-dark distal region of chromosome 21 is responsible for all the symptoms characteristic of Down's syndrome, whereas the bright proximal band gives rise only to mild retardation in the trisomic, and even in the monosomic, state (for example, Hagemeijer and Smit, 1977).

Similarly, it was found that in 20 cases of trisomy for 11q the symptoms were the same whether almost the whole 11q was trisomic or only the distal end (Pihko et al, 1981). The inevitable conclusion was that the genes causing the trisomy symptoms are concentrated in the region distal to 11q23.

Such observations are reflected in the replication pattern of the different types of bands. The Q-dark (G-light) bands replicate during early S period, whereas the Q-bright (G-dark) bands, C-bands, and the inactive X replicate during the latter part of S. The two replication phases do not overlap; indeed, there seems to be a pause between them (cf. Schmidt, 1980; Camargo and Cervenka, 1982). The early replicating bands presumably contain the active genes, whereas the late replicating chromosome regions are not transcribed (cf. Goldman et al, 1984).

References

Boué J, Daketsé M-J, Deluchat G, et al (1976) Identification par les bandes Q et G des anomalies chromosomiques dans les avortements spontanés. Ann Genet 19: 233–239

Brown SW (1966) Heterochromatin. Science 151: 417–425

Camargo M, Cervenka J (1982) Patterns of DNA replication of human chromosomes. II. Replication map and replication model. Am J Hum Genet 34: 757–780

Darlington CD, La Cour L (1940) Nucleic acid starvation of chromosomes in *Trillium*. J Genet 40: 185–213

Doolittle WF (1982) Selfish DNA after fourteen months. In: Dover GA, Flavell RB (eds) Genome evolution. Academic, New York, pp 3–28

Evans HJ (1977) Some facts and fancies relating to chromosome structure in man. In: Harris H, Hirschhorn K (eds) Advances in human genetics, Vol 8. Plenum, New York, pp 347–438

Finnegan DJ, Will BH, Bayev AA, et al (1982) Transposable DNA sequences in eukaryotes. In: Dover GA, Flavell RB (eds) Genome evolution. Academic, New York, pp 29–40

Flavell RB (1982) Sequence amplification, deletion and rearrangement: Major sources of variation during species divergence. In: Dover GA, Flavell RB (eds) Genome evolution. Academic, New York, pp 301–323

Gall JG (1981) Chromosome structure and the C-value paradox. J Cell Biol 91: 3s–14s

Goldman MA, Holmquist GP, Gray MC, et al (1984) Replication timing of genes and middle repetitive sequences. Science 224: 686–692

Goodpasture C, Bloom SE (1975) Visualization of nucleolar organizer regions in mammalian chromosomes using silver staining. Chromosoma 53: 37–50

Hagemeijer A, Smit EME (1977) Partial trisomy 21. Further evidence that trisomy of band 21q22 is essential for Down's phenotype. Hum Genet 38: 15–23

Heitz E (1931) Die Ursache der gesetzmässigen Zahl, Lage, Form und Grösse pflanzlicher Nukleolen. Planta 12: 775–844.
Heitz E (1933) Die Herkunft der Chromocentren. Planta 18: 571–635.
Henderson AS, Warburton D, Atwood KC (1972) Location of ribosomal DNA in the human chromosome complement. Proc Natl Acad Sci USA 69: 3394–3398
Hood L, Campbell JH, Elgin SCR (1975) The organization, expression, and evolution of antibody genes and other multigene families. Annu Rev Genet 9: 305–353
Hsu TC (1975) A possible function of constitutive heterochromatin: the bodyguard hypothesis. Genetics 79: 137–150
Jones RN, Rees H (1982) B chromosomes. Academic, New York
Korenberg JR, Engels WR (1978) Base ratio, DNA content, and quinacrine-brightness of human chromosomes. Proc Natl Acad Sci USA 75: 3382–3386
Korenberg JR, Therman E, Denniston C (1978) Hot spots and functional organization of human chromosomes. Hum Genet 43: 13–22
Kuhn EM (1976) Localization by Q-banding of mitotic chiasmata in cases of Bloom's syndrome. Chromosoma 57: 1–11
Kuhn EM, Therman E, Denniston C (1985) Mitotic chiasmata, gene density, and oncogenes. Hum Genet 70: 1–5
Miklos GLG (1982) Sequencing and manipulating highly repeated DNA. In: Dover GA, Flavell RB (eds) Genome evolution. Academic, New York, pp 41–68
Miklos GLG, John B (1979) Heterochromatin and satellite DNA in man: properties and prospects. Am J Hum Genet 31: 264–280
Miller OJ (1981) Nucleolar organizers in mammalian cells. In: Bennett MD, Bobrow M, Hewitt G (eds) Chromosomes Today, Vol. 7. Allen and Unwin, London, pp 64–73
Nagl W (1982) Condensed chromatin: Species-specificity, tissue-specificity, and cell cycle-specificity, as monitored by scanning cytometry.In: Nicolini C (ed) Cell growth. Plenum, New York, pp 171–218
Passarge E (1979) Emil Heitz and the concept of heterochromatin: longitudinal chromosome differentiation was recognized fifty years ago. Am J Hum Genet 31: 106–115
Patau K (1973) Three main classes of constitutive heterochromatin in man: intercalary, Y-type and centric. In: Wahrman J, Lewis KR (eds) Chromosomes today, Vol 4. Wiley, New York, p 430
Pihko H, Therman E, Uchida IA (1981) Partial 11q trisomy syndrome. Hum Genet 58: 129–134
Ris H, Korenberg JR (1979) Chromosome structure and levels of chromosome organization. In: Goldstein L, Prescott DM (eds) Cell biology. Academic, New York, pp 268–361
Schmidt M (1980) Two phases of DNA replication in human cells. Chromosoma 76: 101–110
Schnedl W, Abraham R, Dann O, et al (1981) Preferential fluorescent staining of heterochromatic regions in human chromosomes 9, 15, and the Y by D287/170. Hum Genet 59: 10–13
Vanderlyn L (1949) The heterochromatin problem in cyto-genetics as related to other branches of investigation. Bot Rev 15: 507–582
Vosa CG (1976) Heterochromatin classification in Vicia faba and Scilla sibirica. In: Pearson PL, Lewis KR (eds) Chromosomes today, Vol 5. Wiley, New York, pp 185–192
Weisblum B (1973) Fluorescent probes of chromosomal DNA structure: three classes of acridines. Cold Spring Harbor Symp Quant Biol 38: 441–449
Weisblum B, de Haseth PL (1972) Quinacrine, a chromosome stain specific for

deoxyadenylate-deoxythymidylate-rich regions in DNA. Proc Natl Acad Sci USA 69: 629–632

Wilson EB (1925) The cell in development and heredity, 3rd edn. MacMillan, New York

Yamasaki N (1956) Differentielle Färbung der somatischen Metaphasechromosomen von *Cypripedium debile*. Chromosoma 7: 620–626

Yunis JJ (1965) Interphase deoxyribonucleic acid condensation, late deoxyribonucleic acid replication, and gene inactivation. Nature 205: 311–312

Yunis JJ, Prakash O (1982) The origin of man: chromosomal pictorial legacy. Science 215: 1525–1530

VI

Fine Structure and Function of the Eukaryotic Chromosome

Chromosome Fine Structure

Our understanding—still far from complete—of the fine structure and function of a eukaryotic chromosome has taken shape during the last decade. The results have been reviewed in numerous books and articles (for example, Lewin, 1980; Gall, 1981; Alberts et al, 1983).

A eukaryotic chromosome consists of *chromatin*, which is a combination of DNA and proteins, mainly histones. One chromatid in a human chromosome of about 5μ length represents a single continuous DNA double helix which, when stretched out, would be 5 cm long. One of the main questions concerning chromosome structure is how a 5 cm-long strand is packaged into a 5μ-long chromatid (cf. Gall, 1981).

At the first level of packaging, histones, of which a eukaryotic chromosome contains five types (H1, H2A, H2B, H3 and H4), are essential components. This fundamental role of the histones is also reflected in their conservation throughout the higher organisms.

Figure VI.1 illustrates the current ideas on chromosome structure. The first stage in the packaging is provided by *nucleosomes*. A nucleosome is a structure combining DNA and histones. Two molecules of each of the histones H2A, H2B, H3, and H4 form an octamer, which is a flat disk around which the DNA double helix is wound twice. Such a structure is called a *nucleosome core*, and the cores are attached to each other by DNA *linkers* (Fig. VI.1, second from top). A nucleosome thus consists of the core and the linker, and the DNA involved in it contains 200 base pairs (cf. Kornberg and Klug, 1981). Histone 1 attaches the nucleosomes to each other, the strand now being 30 nm (nanometers) thick. When chromatin is unravelled, the chromosome strands appear in electron

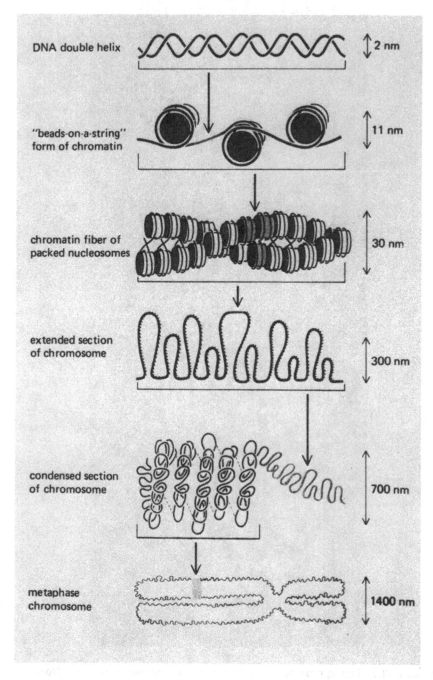

DNA double helix ‖ 2 nm

"beads-on-a-string" form of chromatin ‖ 11 nm

chromatin fiber of packed nucleosomes ‖ 30 nm

extended section of chromosome ‖ 300 nm

condensed section of chromosome ‖ 700 nm

metaphase chromosome ‖ 1400 nm

Figure VI.1. Diagram of the different orders of chromatin packing assumed to give rise to a metaphase chromosome (Fig. 8–24 from Alberts et al, Molecular Biology of the Cell, 1983).

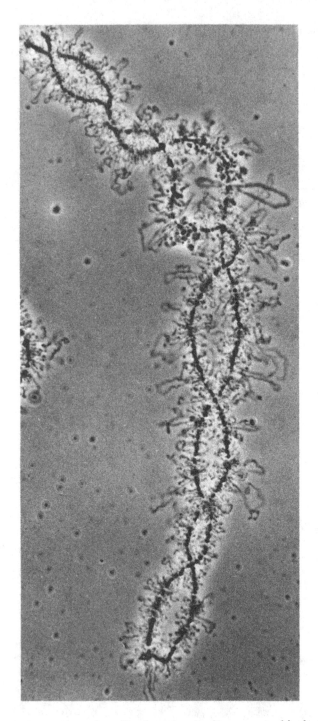

Figure VI.2. The appearance of a lampbrush chromosome bivalent from an oocyte of the salamander *Pleurodeles waltlii* (phase contrast). × 800 (courtesy of U. Scheer).

micrographs as strings with spherical disk-like beads (Fig. VI.1). Interestingly, purified histones will form nucleosomes with any DNA molecule, including bacterial DNA, which normally is not associated with histones. When a gene is transcribed, the chromatin is extended.

Nucleosome formation shortens the DNA double helix some 6–7 times, and it is now 30 nm thick. The next packing order is provided by *domains*, which are folded loops in the 30 nm strand, which now will be 300 nm thick (Fig. VI.1). One domain may contain one or more transcription units or, in other words, genes. The looped domains are assumed to correspond to the loops in lampbrush chromosomes, described below (Fig. VI.2). This assumption is further supported by the appearance of a metaphase chromosome from which the ribonucleoprotein coat has been removed, since it closely resembles a lampbrush chromosome (cf. Miller, 1981).

The 300 nm strand, which has the looped domains (Fig. VI.1), forms a chromatid spiral, which sometimes is visible in the light microscope (Fig. II.1). Apparently the spiral is tighter in some regions than others; this may reflect different types of bands, for instance C-bands, in the metaphase chromosome.

When a chromosome is extended, as during the meiotic prophase stages leptotene, zygotene, and pachytene (Fig. XIV.4), it appears to consist of different-sized condensed *chromomeres* and less condensed regions between them. Presumably the chromomeres represent folded groups of looped domains (cf. Ris and Korenberg, 1979).

Polytene and Lampbrush Chromosomes

Our understanding of the fine structure and function of eukaryotic chromosomes has been greatly advanced through studies on certain specialized chromosome types in which some features are enormously magnified and therefore amenable to analysis.

Polytene chromosomes occur in various types of cells in protozoa, plants, and animals, including mammals (Chapters XII, XIII). They are formed by extended chromosome strands that have replicated repeatedly. Especially informative have been the larval dipteran polytene chromosomes, in which the DNA strands are tightly paired.

A dipteran polytene chromosome is seen to consist of darker bands of variable thickness and lighter interband regions (Fig. XII.4c). The bands are made up of aligned folded domains and correspond roughly to a gene. A polytene chromosome in a *Drosophila* salivary gland shows some 5,000 bands. In the interband regions, whose function is unknown, the DNA strands are extended.

When a gene becomes active, the folded DNA in a band unravels forming a *puff*, which also contains RNA, and when the activity ceases, the

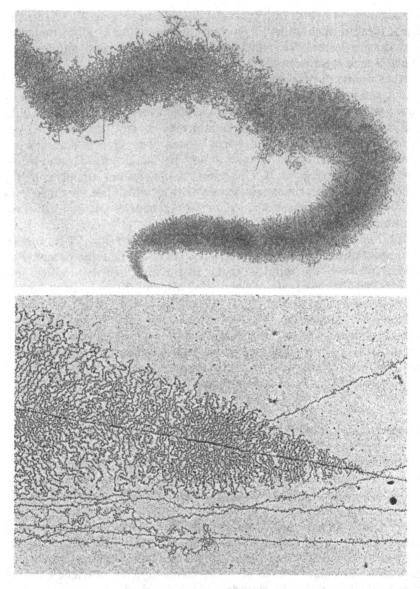

Figure VI.3. Top: low power electron micrograph of a lampbrush chromosome from *Pleurodeles waltlii*. The central axis is covered with nascent ribonucleoprotein fibrils; the uncovered part is the start of the transcription unit. × 5,000 (courtesy of U. Scheer). Bottom: same preparation as in the above micrograph, showing the start region of a transcribed unit. The densely packed RNA polymerase granules are clearly visible in the central axis. The strands not covered with fibrils represent transcriptionally inactive chromatin showing nucleosomes as little dots. × 20,000 (courtesy of U. Scheer).

puff recedes, changing back into a normal band. A puff labels heavily with ^{3}H-uridine; this demonstrates active RNA synthesis. At different stages of the insect's development, different bands are puffed.

The other type of specialized chromosome that has been especially informative in studies of gene action is a *lampbrush chromosome*. These are found in the growing oocytes of a variety of vertebrates and invertebrates, and even in the primary nucleus of the green alga *Acetabularia* (cf. Scheer et al, 1979). Indeed, the hypothesis has been proposed that all eukaryotic chromosomes resemble lampbrush chromosomes in their basic plan. The most intensively studied lampbrush chromosomes are those found in the giant oocytes of amphibia.

Lampbrush chromosomes represent a special type of diplotene chromosome. The bivalents are enormously elongated, and from the chromomeres extend pairs of DNA loops, giving the chromosomes their lampbrush appearance (Fig. VI.2). In cross section such a chromosome is star-like, with loops projecting from a dense chromomere core (cf. Scheer et al, 1979).

Intensive synthetic activity goes on in the loops; it has been estimated that a thousand genes in a cell may be transcribed simultaneously. As a result the loops are covered with ribonucleoprotein fibrils, which make them thicker and therefore visible in a phase contrast microscope (Fig. VI.3). The RNA fibrils are attached to the loop axis via RNA polymerase-containing granules, and the lengths of the fibrils form continuous gradients from the starting point of the transcription (Fig. VI.3). Each loop may contain one or more transcriptional units.

At specific loops, DNA rings—apparently copies of the loops—are synthesized and extruded into the nucleoplasm. They replicate extrachromosomally through a rolling circle mechanism. In the nucleoplasm the rings form additional nucleoli that may number 200–300 in an amphibian oocyte. In human diplotene oocytes some 20–30 extra nucleoli are found (Stahl et al, 1975). These processes reflect the enormous growth of the oocytes at these stages.

References

Alberts B, Bray D, Lewis J, et al (1983) Molecular biology of the cell. Garland, New York

Gall JG (1981) Chromosome structure and the C-value paradox. J Cell Biol 91: 3s–14s

Kornberg RD, Klug A (1981) The nucleosome. Sci Am 244: 52–64

Lewin B (1980) Gene expression 2 (2nd edition). Wiley, New York

Miller OL Jr (1981) The nucleolus, chromosomes, and visualization of genetic activity. J Cell Biol 91: 15s–27s

Ris H, Korenberg JR (1979) Chromosome structure and levels of chromosome organization. In: Goldstein L, Prescott DM (eds) Cell biology. Academic, New York, pp 268–361

Scheer U, Spring H, Trendelenburg MF (1979) Organization of transcriptionally active chromatin in lampbrush chromosome loops. In: Busch H (ed) The cell nucleus, Vol 7. Academic, New York, pp 3–47

Stahl A, Luciani JU, Devictor U, et al (1975) Constitutive heterochromatin and micronucleoli in the human oocyte at the diplotene stage. Humangenetik 26: 315–327

VII
Chromosome Structural Aberrations

Origin of Structurally Abnormal Chromosomes

Chromosomes sometimes break spontaneously, or breakage may be caused by a mutagenic agent, such as ionizing radiation or a chemical compound. Unlike normal chromosome ends, broken ends tend to join each other. Usually the broken ends rejoin; in other words, the break heals. However, a break may lead to a deletion or, if more than one break has occurred in a cell, to structural rearrangements of chromosomes. At least three different DNA repair systems may be involved in the joining of broken chromosome ends (cf. Bartram, 1980).

Chromosomes may break at any stage of the cell cycle—G_1, S, G_2— during mitosis or during meiosis. Various cell types and stages show very different responses to chromosome-breaking agents even in the same organism. Thus, in *Vicia*, a dose of x-rays that causes approximately one aberration per cell in G_2 induces one aberration visible at metaphase per 10 cells when irradiation is given in G_1 (cf. Evans, 1974). Even greater differences are found when different plant and animal species are compared. When higher plants were tested for the dose of radiation needed to inhibit their growth or to kill them, the differences were more than 100-fold (Sparrow, 1965).

The study of chromosome breaks is intimately connected with research on gene mutations, because most mutagens induce both chromosome breaks and gene mutations. Often the same agents are also carcinogenic. We know that the radiations and drugs used in cancer therapy can also cause malignant disease. Various aspects of chromosome breakage are reviewed in numerous books and articles, of which the following reflect a somewhat arbitrary sample: Kihlman, 1966; Rieger and Michaelis,

1967; Gebhart, 1970; Evans, 1962, 1974, 1983; Auerbach, 1976, 1978.

Chromosome Breaks and Rearrangements

If a chromosome breaks during the G_1 stage, when it consists of only one chromatid, the break will be perpetuated in S and will affect *both* chromatids in the following metaphase. A single break may either rejoin or result in a deleted chromosome and an acentric fragment (Fig. VII.1b) that will be lost in a subsequent mitosis, or the acentric fragment is included in a daughter nucleus and replicates, so that there will be double fragments in the next metaphase.

Two breaks in the same chromosome may result in the formation of

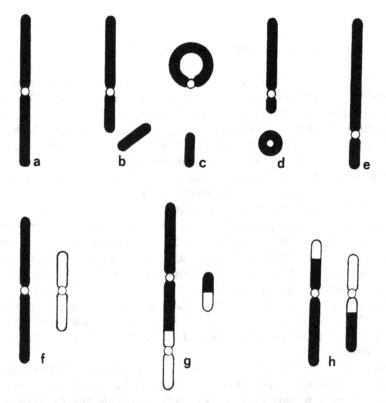

Figure VII.1. Results of G_1 breaks in one chromosome (a), and in two chromosomes (f); (b) broken chromosome; (c) centric ring and acentric fragment; (d) acentric ring and centric fragment; (e) chromosome with pericentric inversion; (g) dicentric chromosome and acentric fragment; (h) balanced reciprocal translocation.

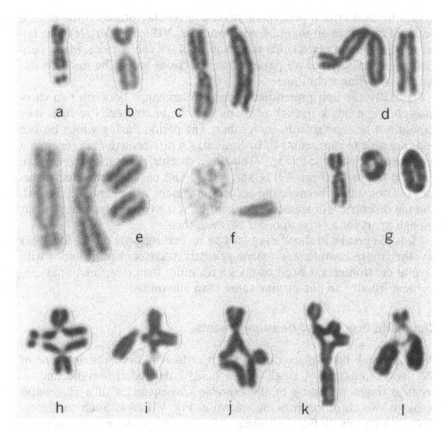

Figure VII.2. Chromosome structural abnormalities. (a) Gap; (b) gap through centromere; (c) normal chromosome 1 and its homolog with a pericentric inversion; (d) dicentric chromosome and acentric fragment; (e) two dicentrics and two acentrics from the same cell (courtesy of EM Kuhn); (f) an interphasic D chromosome in satellite association with a normal D; (g) chromosome 9, ring(9) and double ring(9) (courtesy of ML Motl); (h) mitotic chiasma between heteromorphic homologs; (i) class II quadriradial between two D chromosomes in satellite association with a D and a G; (j–k) class IVb chromatid translocations; (l) hexaradial chromatid translocation or a satellite association between two D chromosomes and a G.

either a centric ring and an acentric fragment or an acentric ring and an interstitial deletion (Figs. VII.1c and d and Fig. VII.2g). A segment that is deleted interstitially from a chromosome arm may remain as an acentric fragment if its ends fail to join. Very small fragments are called minutes.

If the breaks take place in the same arm, another possible result of a two-break intrachange is a *paracentric* inversion, in which the deleted segment rejoins the chromosome in an inverted position. If one break occurs in each arm, a *pericentric* inversion will be formed if the deleted segment rejoins in an inverted position (Figs. VII.1e and VII.2c). The latter rearrangement often shifts the position of the centromere. Many pericentric inversions and all paracentric inversions would be undetectable without banding techniques.

Naturally the first prerequisite for an interchange between two chromosomes is a break in each of them. For the broken ends to fuse, they should not be too far from each other. The period during which broken ends are able to join seems to be limited. An interchange may result in a reciprocal translocation (Fig. VII.1h) or a dicentric chromosome and an acentric fragment (Figs. VII.1g and VII.2d and e). The latter rearrangement is not stable, because the acentric fragment will eventually be lost, and the dicentric will encounter difficulties if the two centromeres are far enough apart for a twist to occur between them.

Multiple breaks in a cell may lead to several reciprocal translocations or other more complicated rearrangements, such as chromosomes with several centromeres. Chromosomes with more than one functional centromere usually do not survive more than one mitosis.

Chromatid Breaks and Rearrangements

When a break takes place during G_2, it ordinarily involves only one of the two chromatids. A single break yields a deleted chromatid and an acentric fragment. Some of the possible consequences of a chromatid break in two chromosomes are shown in Fig. VII.2i–k. Such configurations, which result from chromatid exchanges between two chromosomes, are called *quadriradials*. They come in two types, depending on whether the centromeres are on opposite sides (alternate) (I, IIIa, and IVa in Fig. VII.3), or next to each other (adjacent) (II, IIIb, and IVb in Fig. VII.3). The *alternate* type leads to the formation of two chromatids, each with a reciprocal translocation and two unchanged chromatids. The *adjacent* arrangement gives rise to a dicentric and an acentric chromatid and two unchanged chromatids. Mitotic chiasmata (I in Fig. VII.3), which form a special subgroup of the alternate type of quadriradials, are discussed in Chapter X.

Telomeres

Unbroken chromosome ends do not as a rule show any tendency to join each other. They are assumed to be capped by so-called telomeres. A long-standing dispute is: Can a broken chromosome end heal and thereafter act as a telomere, or is a "real" telomere needed to make the end normal again?

There is evidence, at least in the human, that broken chromosome ends

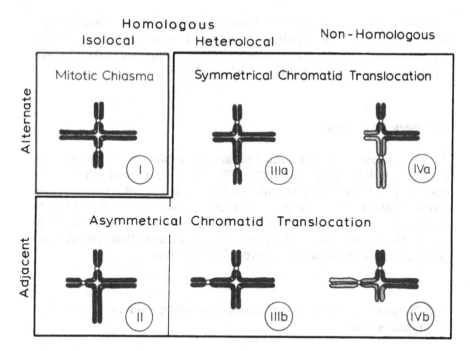

Figure VII.3. The classification of quadriradial configurations (Therman and Kuhn, 1976).

are able to heal (Patau, 1965). (1) If a chromosome is broken with x-rays, the sister chromatids do not show a noticeable tendency to join, unlike the behavior of some plant chromosomes. (2) Cases of terminally deleted chromosomes are too frequent and the breakpoints too consistent for them to be the result of two breaks. (3) In the cri du chat syndrome, which is caused by the deletion of about one-half of the short arm of chromosome 5, the broken ends may appear fuzzy, thus differing from normal chromosome ends (Patau, 1965). In addition, Niebuhr (1978) showed that in 35 cri du chat patients the deletion appeared terminal in 27 of them, interstitial in four, and capped by a reciprocal translocation in four others. (4) Ring chromosomes may sometimes open up and act like normal two-armed chromosomes (for example, Cooke and Gordon, 1965).

However, it was recently shown with NOR-staining in two cases of translocation between the short arm of acrocentrics and other autosomes that the nucleolar organizing region of the short arm was attached to the broken end of the other translocation chromosome (Hansmann et al, 1977). Similarly, in two human meningiomas the broken ends of chromosomes 1 and 16 were capped by NOR-stained nucleolar organizer material (Zankl and Huwer, 1978). Such observations indicate that telomeres may be needed to cap the broken ends, at least for some breakpoints.

The molecular structure of telomeres has lately been studied intensively, and most of the hypotheses presented assume that telomeres have a palindromic structure which allows the DNA to fold back like a hairpin (cf. Bateman, 1975; Holmquist and Dancis, 1979).

Telomere Association

Telomeres seem to have a mutual attraction, forming reversible or stable associations (cf. Drets and Stoll, 1974). Dutrillaux et al (1977) have described an interesting phenomenon in the Thiberge-Weissenbach syndrome. Unbroken chromosome ends tend to join each other, leading to the formation of chains and rings, sometimes involving all the chromosomes. No acentric fragments were present, indicating that this phenomenon cannot be the result of reciprocal translocations.

The Origin of Dicentric Chromosomes Including Isodicentrics and Isochromosomes

Isodicentric chromosomes are symmetric, palindromic structures consisting of two homologous chromosomes broken at identical points. Usually one centromere is inactivated. Isochromosomes, on the other hand, which are either monocentric or dicentric, are metacentrics having two identical arms. A dicentric isochromosome corresponds to an isodicentric in which the centromeres are next to each other. Most isodicentrics have been formed by two X chromosomes (Fig. XX.4b, c) (Zakharov and Baranovskaya, 1983), and most isochromosomes by X long arms. Isodicentrics consisting of two Y chromosomes have been found repeatedly (cf. Cohen et al, 1973). Dicentrics between various nonhomologous chromosomes, with one centromere inactivated, have also been observed (cf. Schwartz et al, 1983), as have many Robertsonian translocations with two centromeres.

Figure VII.4 illustrates two possible modes of origin of isodicentrics and dicentric isochromosomes. (A G_1 break at identical points in two homologous chromosomes is too unlikely an event to need consideration). In Fig. VII.4a both chromatids of a chromosome are broken and have joined. However, if the breakpoints are at any distance from the centromere, this mechanism would not lead to an isodicentric, but a bridge in next anaphase, since the daughter centromeres would go to opposite poles. If the daughter centromeres are very near each other, they are more likely to go to the same pole which would give rise to a dicentric isochromosome.

The most probable origin of isodicentric chromosomes, including

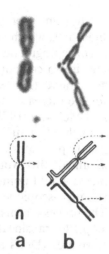

Figure VII.4. Possible origins of isodicentric chromosomes. (a) Isochromatid break and rejoining of the broken chromatids rarely results in an isodicentric; (b) segregation of an adjacent quadriradial is the main mechanism creating isodicentrics.

a   b

dicentric isochromosomes, is *segregation of an adjacent quadriradial* (type II in Fig. VII.3) when the centromeres of the dicentric chromatid go to the same pole (Fig. VII.4b). Therman and Kuhn (1985) have shown that this is the usual mechanism for creating symmetric dicentrics in Bloom's syndrome. Dicentrics between two nonhomologous chromosomes arise through segregation of a type IVb quadriradial (Fig. VII.3), or a G_1 break in two nonhomologous chromosomes.

A quadriradial may segregate in different ways. However, common to all of them is that the daughter cells are *different from each other*, whereas descendants of a cell in which a G_1 aberration has taken place are identical. Many human mosaics display cell lines with different chromosome constitutions that obviously owe their origin to segregation in a quadriradial (cf. Daly et al, 1977). Good examples of such mosaics are provided by persons having a cell line with an isodicentric X;X chromosome and another with 45,X chromosomes. Although segregation in a quadriradial is the simplest explanation for such mosaics, it has been largely neglected in the literature.

Inactivation of the Centromere

Dicentric chromosomes with any distance between the centromeres will survive only if one centromere is inactivated. This is what has happened in most isodicentrics and dicentrics between homologous chromosomes. Such a chromosome functions like a normal monocentric (Therman et al, 1974). No second primary constriction is visible, but a C-band marks its position (Fig. XX.4b, c).

The same process obviously takes place in monocentric human X chromosomes, which have a nonfunctional centromere and the appearance of acentric fragments. Such chromosomes drift at random in anaphase, giving rise to cells in which one X is missing and to others in which they accumulate (Fig. XX.4a). This obviously is the mechanism that accounts for the increased frequency of 45,X cells in older women (cf. Fitzgerald and McEwan, 1977).

In the "acentric" X chromosomes the centromere certainly does not divide prematurely (cf. Nakagome et al, 1984), as has sometimes been assumed; this would create a different configuration, since the centromeres would point towards the poles. However, what is involved at the submicroscopic level in the inactivation of a centromere is not known. The process seems to be irreversible, as demonstrated by the behavior of a 6p;19p translocation, in which the chromosomes were attached at their telomeres (Drets and Therman, 1983) and the centromere of chromosome 6 was inactivated. This chromosome had a tendency to break at the fusion point, and in none of the "acentric" chromosomes 6 was the centromere reactivated (Fig. XXII.7).

Misdivision

Isochromosomes with one centromere and telocentrics arise through *misdivision of the centromere*. This means that in mitotic or meiotic anaphase the centromere divides transversely instead of longitudinally (Fig. XVI.2). Since misdivision most often takes place in the first meiotic anaphase, it will be discussed in Chapter XVI.

Centric Fusion

Whole-arm transfers, or Robertsonian translocations as they are called, constitute a special class of reciprocal translocations. In the human they almost always occur between two acrocentric chromosomes and are the most commonly observed chromosome aberrations.

The origin of Robertsonian translocations has often been assumed to be the fusion of two centromeres that have each broken in the middle. Recent banding studies show that this is not the only, and possibly not even the most common, mechanism. As illustrated in Fig. VII.5, depending on whether the break in the two acrocentrics occurs through the centromere or on the short or the long arm, the result is a monocentric or dicentric (or acentric) translocation chromosome, consisting of two long arms. The reciprocal product, made up of two short arms, is inevitably lost if it does not have a centromere. In fact, even when such a chromosome possesses a centromere, it is usually lost because it consists entirely of heterochromatin. Neither the absence of these small chromosomes nor their presence as extra units seems to affect the phenotype.

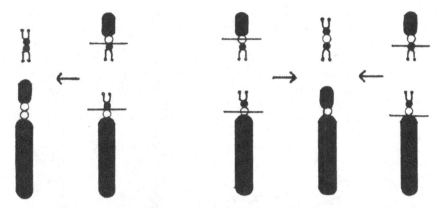

Figure VII.5. Origin of Robertsonian translocations between a G and a D chromosome. Left: Breaks on the short arms resulting in a dicentric and acentric chromosome. Right: Break either through the centromeres, or one break on the short arm, the other on the long arm, resulting in two monocentric chromosomes.

Triradial and Multiradial Chromosomes

Triradial and multiradial chromosome configurations deserve only brief mention because of their extreme rarity and are described here only as examples of the odd ways in which chromosomes may behave. In a triradial, part of the chromosome is duplicated; it has three branches in contrast to a quadriradial, which has four. Triradials are much rarer than quadriradials; for instance, Stahl-Maugé et al (1978) found two triradials in 53,000 cells of normal persons. However, they are more frequent in cells of patients with Fanconi's anemia and Bloom's syndrome and in cells treated with chromosome-breaking substances. The branchpoint in a triradial is often a fragile region.

Three different mechanisms have been proposed as creating triradial chromosomes: partial endoreduplication (a segment of a chromosome replicates twice while the rest replicates once), a chromatid fragment that remains associated with its sister chromatid in anaphase, or a broken chromosome inserted into a gap formed by a chromatid break (Stahl-Maugé et al, 1978).The last mechanism will hardly ever create symmetrical triradials. That the second mechanism works at least sometimes has been shown by studies on differentially stained sister chromatids (cf. Weitkamp et al, 1978). However, the following observations show that most symmetric triradials probably arise through partial endoreduplication (Fig. VII.6e) (Kuhn and Therman, 1982). Sometimes the extra segment joins its intact sister chromatid. It is difficult to imagine duplicate satellites coming about through any other process. The extra segments in a triradial are paired with their sister chromatids and not with each other. Finally, x-rays in G_2 increase the frequency of chromatid fragments but not of symmetric triradials.

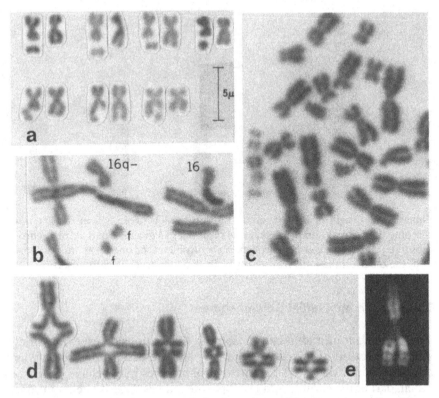

Figure VII.6. Chromosome 16 with a fragile region and its normal homolog from 7 cells; (b) break at the fragile region and the replicated fragment (f); (c) banded C chromosome left behind in its cycle; (d) mitotic chiasmata involving different human chromsome pairs; (e) triradial chromosome 1 resulting from partial endoreduplication.

Multibranched chromosomes have so far been found in the lymphocytes of three patients with combined immunodeficiency (cf. Tiepolo et al, 1979; Fryns et al, 1981). These peculiar configurations consist of variable numbers and combinations of short and long arms of chromosomes 1, 9, and 16, in which the exchanges occur at the centric regions.

Fragile Regions

Fragile sites appear as unstained and often stretched regions in a chromosome (Figs. VII.6.a–b and XX.4d). The chromosomes have a tendency to break at these points and the fragile regions act as branchpoints in triradial chromosomes. They are a constant feature of the region involved, but are expressed in only a proportion of cells. Fragile sites are inherited in a mendelian fashion. They have been useful as markers in chromosomes, especially in gene mapping studies.

Although fragile regions have been subjected to intensive study, both their fine structure and function are unknown. The only fragile site which has a definite phenotypic effect is Xq27 (Fig. XX.4d), whose presence is correlated with a mental retardation syndrome. However, it is interesting that the autosomal fragile sites seem to be more frequent in mental retardates than in newborns (Sutherland, 1982).

Apart from an undefined site Xq26, the fragile regions seem to fall into four groups (cf. de la Chapelle and Berger, 1984).

1. The folate sensitive sites 2q11.2, 2q13, 6p23, 7p11.2, 8q22.3, 9p21.1, 9q32, 10q23.3, 11q13.3, 11q23.3, 12q13.1, 16p12.3, 20p11.23, and Xq27.3 can be induced with low levels of folic acid and thymidine in the culture medium.

2. Sites 16q22.1 and 17p12 are inducible with distamycin A, but also appear spontaneously.

3. The fragile site 10q25.2 is brought about with BrdU.

4. The sites 3p14, 6q26, and 16q23 occur commonly, but are also induced with low folate levels and aphidicolin.

Recently attention has been paid to the possible correlation of fragile sites and constant chromosome breakpoints in cancer (cf. Yunis, 1983; Hecht and Sutherland, 1984).

References

Auerbach C (1976) Mutation research. Problems, results and perspectives. Chapman and Hall, London

Auerbach C (1978) Forty years of mutation research: a pilgrim's progress. Heredity 40: 177–187

Bartram CR (1980) DNA repair: pathways and defects. Eur J Pediatr 135: 121–128

Bateman AJ (1975) Simplification of palindromic telomere theory. Nature 253: 379–380

Chapelle A de la, Berger R (1984) Report of the committee on the chromosome rearrangements in neoplasia and on fragile sites. Cytogenet Cell Genet 37: 274–311

Cohen MM, MacGillivray MH, Capraro VJ, et al (1973) Human dicentric Y chromosomes. J Med Genet 10: 74–79

Cooke P, Gordon RR (1965) Cytological studies on a human ring chromosome. Ann Hum Genet 29: 147–150

Daly RF, Patau K, Therman E, et al (1977) Structure and Barr body formation of an Xp+ chromosome with two inactivation centers. Am J Hum Genet 29: 83–93

Drets ME, Stoll M (1974) C-banding and non-homologous associations in *Gryllus argentinus*. Chromosoma 48: 367–390

Drets ME, Therman E (1983) Human telomeric 6;19 translocation chromosome with a tendency to break at the fusion point. Chromosoma 88: 139–144

Dutrillaux B, Aurias A, Couturier J, et al (1977) Multiple telomeric fusions and chain configurations in human somatic chromosomes. In: Chapelle A de la, Sorsa M (eds) Chromosomes today, Vol 6. Elsevier/North Holland, Amsterdam, pp 37–44

Evans HJ (1962) Chromosome aberrations induced by ionizing radiations. Int Rev Cytol 13: 221–321

Evans HJ (1974) Effects of ionizing radiation on mammalian chromosomes. In: German J (ed) Chromosomes and cancer. Wiley, New York, pp 191–237

Evans HJ (1983) Effects on chromosomes of carcinogenic rays and chemicals. In: German J (ed) Chromosome mutation and neoplasia. Liss, New York, pp. 253–279

Fitzgerald PH, McEwan CM (1977) Total aneuploidy and age-related sex chromosome aneuploidy in cultured lymphocytes of normal men and women. Hum Genet 39: 329–337

Fryns JP, Azou M, Jaeken J, et al (1981) Centromeric instability of chromosomes 1, 9 and 16 associated with combined immunodeficiency. Hum Genet 57: 108–110

Gebhart E (1970) The treatment of human chromosomes in vitro: results. In: Vogel F, Röhrborn G (eds) Chemical mutagenesis in mammals and man. Springer, New York, pp 367–382

Hansmann I, Wiedeking C, Grimm T, et al (1977) Reciprocal or nonreciprocal human chromosome translocations? The identification of reciprocal translocations by silver staining. Hum Genet 38: 1–5

Hecht F, Sutherland GR (1984) Fragile sites and cancer breakpoints. Cancer Genet Cytogenet 12: 179–181

Holmquist GP, Dancis B (1979) Telomere replication, kinetochore organizers, and satellite DNA evolution. Proc Natl Sci USA 76: 4566–4570

Kihlman BA (1966) Actions of chemicals on dividing cells. Prentice-Hall, Englewood Cliffs, New Jersey

Kuhn EM, Therman E (1982) Origin of symmetrical triradial chromosomes in human cells. Chromosoma 86: 673–681

Nakagome Y, Abe T, Misawa S, et al (1984) The "loss" of centromeres from chromosomes of aged women. Am J Hum Genet 36: 398–404

Niebuhr E (1978) Cytologic observations in 35 individuals with a 5p- karyotype. Hum Genet 42: 143–156

Patau K (1965) The chromosomes. In: Birth defects: original article series, Vol 1, The National Foundation—March of Dimes, New York, pp 71–74

Rieger R, Michaelis A (1967) Die Chromosomenmutationen. Gustav Fischer, Jena

Schwartz S, Palmer CG, Weaver DD, et al (1983) Dicentric chromosome 13 and centromere inactivation. Hum Genet 63: 332–337

Sparrow AH (1965) Comparisons of the tolerances of higher plant species to acute and chronic exposure of ionizing radiation. In: Mechanisms of the dose rate effect of radiation at the genetic and cellular levels. Special suppl. Jpn J Genet 40: 12–37

Stahl-Maugé C, Hager HD, Schroeder TM (1978) The problem of partial endoreduplication. Hum Genet 45: 51–62

Sutherland GR (1982) Heritable fragile sites on human chromosomes. VIII. Preliminary population cytogenetic data on the folic-acid-sensitive fragile sites. Am J Hum Genet 34: 452–458

Therman E, Kuhn EM (1976) Cytological demonstration of mitotic crossing-over in man. Cytogenet Cell Genet 17: 254–267

Therman E. Kuhn EM (1985) Incidence and origin of symmetric and asymmetric dicentrics in Bloom's syndrome. Cancer Genet Cytogenet 15: 293–301

Therman E, Sarto GE, Patau K (1974) Apparently isodicentric but functionally monocentric X chromosome in man. Am J Hum Genet 26: 83–92

Tiepolo L, Maraschio P, Gimelli G, et al (1979) Multibranched chromosomes 1, 9, and 16 in a patient with combined IgA and IgE deficiency. Hum Genet 51: 127–137

Weitkamp LR, Ferguson-Smith MA, Guttormsen SA, et al (1978) The linkage relationships of marker sites on chromosomes no. 2 and 10. Ann Hum Genet 42: 183–189

Yunis JJ (1983) The chromosomal basis of human neoplasia. Science 221: 227–236

Zakharov AF, Baranovskaya LI (1983) X-X chromosome translocations and their karyotype-phenotype correlations. In: Sandberg AA (ed) Cytogenetics of the mammalian X chromosome, Part B: X chromosome anomalies and their clinical manifestations. Liss, New York, pp 261–279

Zankl H, Huwer H (1978) Are NORs easily translocated to deleted chromosomes? Hum Genet 42: 137–142

VIII
Causes of Chromosome Breaks

Spontaneous Chromosome Breaks

Spontaneously broken and rearranged chromosomes are occasionally found in every human being and in every cell culture. Their frequency varies from person to person and from culture to culture. One rule, however, appears to be well-established. The incidence of aneuploid cells is known to increase with age, and similarly chromosome structural changes become more frequent in older persons. In our laboratory an analysis of 2324 cells from persons under 40 years of age gave an average of 0.8 percent of cells with chromosome structural aberrations (excluding gaps), whereas in another sample of about the same size from individuals whose average age was 55.8 years, abnormalities were found in 2.4 percent of cells (Kuhn and Therman, 1979). Interestingly, the sensitivity of lymphocytes to alkylating agents increases significantly from newborns to young adults (mean age 23) to old people (mean age 70) (Bochkov and Kuleshov, 1972).

Although the trends are the same, the actual frequencies of chromosome aberrations found in different laboratories show a wide range of variation. This probably depends mainly on different criteria used for scoring aberrations and also on the culture conditions.

Another indication of an age effect is the fact that tissue cultures of normal human fibroblasts can be grown unchanged through only about 50 passages (approximately one year) (Hayflick and Moorhead, 1961). Thereafter, cell growth suffers and more and more aberrations appear. Finally the normal cell strains are transformed into permanent cell lines with numerically and structurally aberrant chromosome constitutions (cf. Nichols, 1975).

When we talk about "spontaneous" chromosome breaks, it means that we can only guess at the actual causes. The genotype of the individual undoubtedly determines the basic level of aberrations. Physiological degeneration in old age probably accounts for the increase in both non-disjunction and breakage of chromosomes. Exposure to cosmic rays and medical or occupational radiation constitute other sources of chromosome damage. Various drugs, viral infections, and even high fever may be yet other sources of damage.

It is well-known that chemically induced chromosome breaks do not occur at random among cells; in other words, they do not follow a Poisson distribution. The same seems to be true of spontaneous breaks (cf. Schroeder and German, 1974; Therman and Kuhn, 1976). Part of the evidence comes from individuals with complicated chromosome rearrangements and is thus incidental. In a population of chromosomally normal grasshoppers, Coleman (1947) found an individual who was heterozygous for three reciprocal translocations and one inversion. Similarly, White (1963) described a grasshopper with a complex translocation involving breaks in four nonhomologous chromosomes. In man, too, complicated rearrangements are occasionally encountered. For example, in a child with congenital anomalies, four chromosomes showed altogether six breaks (Seabright et al, 1978), and Bijlsma et al (1978), who have reviewed the literature on multiple break cases, report a family in which two reciprocal translocations were segregating.

As discussed in Chapter VII, breaks often occur at the fragile chromosome sites.

Radiation-Induced Breaks

Ultraviolet light and especially various types of ionizing radiation are powerful chromosome-breaking agents. The effects of a wide range of radiations, such as x-rays, gamma-radiation, alpha and beta particles, and neutrons, have been studied in this respect. Ionizing radiation causes chromosome breaks at any stage of the mitotic cycle or during meiosis, although the vulnerability of different phases and different organisms varies greatly. The results also vary according to whether a given dose of radiation is applied within a short time span or over a longer period or is fractionated. The tritium in ^{3}H-thymidine, so widely used for autoradiography, emits beta particles that may cause considerable damage when incorporated into the chromosomes.

An enormous amount of research has been undertaken to elucidate the actual mechanism of radiation-induced chromosome breakage. In the late 1930s the so-called target theory was born. It states that a chromosome is broken when it is hit directly by an ion or an ion cluster. However, because a number of observations cannot be explained on the basis of the target theory, the chemical theory of chromosome breakage was proposed in the 1940s. This theory says that most of the chromosome breakage is done by substances that result from the chemical reactions induced by radiation. The various radicals that arise from irradiated water molecules seem to be especially active in this respect.

The following are some of the observations that do not agree with the target theory but that can be readily explained by the chemical theory. Chromosomes of many plants are broken with much lower doses of radiation than, for instance, *Drosophila* chromosomes. Chromosome breaks can be caused by a preirradiated medium. One of the most important observations is that oxygen greatly enhances the effects of radiation, and that, inversely, a number of reducing substances—especially those containing sulfhydryl groups—counteract radiation damage (cf. Evans, 1974). As has been stressed repeatedly, especially by Auerbach (1976, 1978), reactions caused by radiations—as well as those induced by chemical mutagens—take place in a *living cell*, a fact that in many ways influences the final outcome of any treatment.

Chemically Induced Breaks

The list of substances found to break chromosomes is already very long and is still growing steadily. It includes alkylating agents, nucleic acid analogues, purines, antibiotics, nitroso-compounds, and a large number of miscellaneous substances (cf. Kihlman, 1966; Rieger and Michaelis, 1967; Gebhart, 1970; Auerbach, 1976). Such substances have often been called radiomimetic or clastogenic.

Most chemicals induce breaks of the G_2 type, but they have to be present during the preceding S period. As with radiation treatment, different organisms and tissues show a wide range of responses to the same chemical. An example is illustrated in Fig. VIII.1. Two anticancer drugs of the methyl-benzyl hydrazine group broke no chromosomes either in malignant or normal cells in vitro. In vivo they showed no effect on the chromosomes in normal mouse spleen, bone marrow, or spermatocytes. However, they caused extensive damage in transplantable ascites tumors in the same animal (Therman, 1972; Rapp and Therman, 1977). It seems obvious that the mouse is needed to metabolize and change the drug and that it undergoes a second change in the cancer cells. Apparently at least two steps are needed to transform the methyl-benzyl hydrazines into chromosome-breaking substances. All the breaks are of the G_2 type and take place in the Q-dark or G-light regions (Rapp and Therman, 1977).

Virus-Induced Breaks

The discovery that many viruses induce chromosome breaks is relatively recent. This aspect of chromosome breakage has therefore been studied much less than has radiation or chemical mutagenesis, and very little is

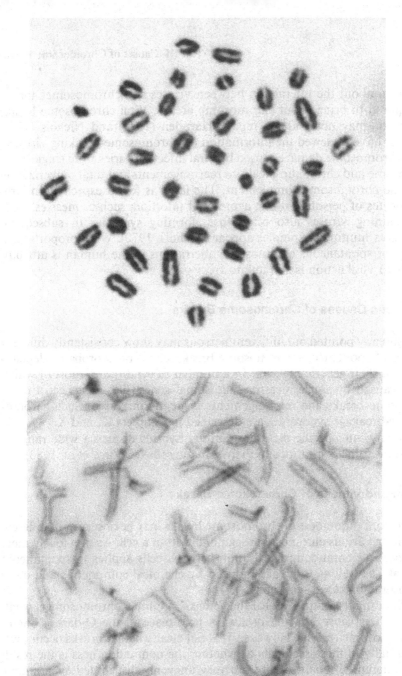

Figure VIII.1. Top: Normal spermatogonial metaphase of the mouse ($2n = 40$). Bottom: Chromosome breaks and numerous chromatid translocations in a mouse cancer cell caused with 1-methyl-2-benzyl hydrazine. (The cancer chromosomes have been spread by adding water to the fixative and therefore stain differently.)

known about the interaction between viruses and chromosomes that is required to bring about chromosome breaks. Viral chromosome breaks may or may not undergo repair. Harnden (1974) and Nichols (1975; 1983) have reviewed the information on chromosome-breaking viruses.

Chromosome damage caused by viral infection varies from single chromosome and chromatid breaks to rearrangements and total pulverization of the chromosome complement. The latter is found especially in lymphocytes of persons with an acute viral infection, such as measles. Cells containing viruses also often fuse, forming syncytia. In subsequent mitoses multipolar spindles appear (Nichols, 1970). What proportion, if any, of spontaneous chromosome aberrations in the human is attributable to viral action is still unknown.

Genetic Causes of Chromosome Breaks

As already pointed out, different persons may show consistently different rates of spontaneous chromosome breakage, the rates probably depending on their genetic makeup. In a number of syndromes, most of which are caused by single recessive autosomal genes, the incidence of chromosome breaks and rearrangements is greatly increased. Such chromosome-breakage syndromes are described in Chapters IX and X.

Many interspecific plant and animal hybrids display a wide range of chromosome aberrations in their somatic cells (cf. Shaw et al, 1983).

Nonrandomness of Chromosome Breaks

The nonrandomness of chromosome breaks may be observed in different cells or various chromosome segments within a cell. As previously mentioned, a nonrandom distribution between cells applies to spontaneous breaks as well as to breaks induced by chemical compounds and chromosome-breakage syndromes.

Banding techniques reveal that breaks in human chromosomes, whatever their cause, practically always take place in the Q-dark (G-light) chromosome regions. However, it is not clear whether breaks occur preferentially in these segments or whether the nonrandomness is the result of differential repair. Breaks are very unevenly distributed even among the Q-dark segments, with certain of them emerging as clear hot spots. But the hot spots are not the same when chromosome breakage caused by different agents is compared (cf. von Koskull and Aula, 1977).

Detailed studies on chromosome breakage by various mutagens have been done by Rieger and his colleagues using the large chromosomes of *Vicia faba* ($2n = 12$) (Rieger and Michaelis, 1972; Rieger et al 1973, 1982). For instance, maleic hydrazide breaks especially certain C-bands (Schubert et al, 1981), and the nucleolus organizing region is sensitive to various mutagens (Schubert and Rieger, 1980). In human chromosomes, mitomycin C induces breaks largely in the heterochromatin of chromosomes 1, 9, 16, and the acrocentrics (Shaw and Cohen, 1965).

Methods in Chromosome-Breakage Studies

In plants and animals, chromosome breakage has been analyzed both in vivo and in vitro. In vivo studies are naturally preferable for many investigations, since they permit us to observe the effects of a mutagen in an intact organism. Effects in vivo often differ drastically from the mutagen's effects on cells in culture.

A recent promising technique in chromosome aberration studies is cell fusion and the resulting formation of prematurely condensed chromosomes (Chapter XI).

In man, in vivo experiments are usually impractical or unethical, or both. Nevertheless,it is possible to analyze chromosome damage in persons who have been exposed to ionizing radiation for medical reasons, because of their occupation, or as a result of an accident. The most extensively studied persons are the atomic bomb survivors of Hiroshima and Nagasaki; results of chromosome studies on them are reviewed by Awa (1974). The victims show a significantly higher incidence of chromosome abnormalities in their lymphocytes than do nonexposed controls. One of the most interesting phenomena is that those who received heavy doses of radiation still exhibit, more than 30 years later, abnormal chromosomes, such as dicentrics, acentrics, reciprocal translocations, inversions, rings, and deleted chromosomes. Often clones of cells with the same abnormal chromosome constitution have been found. The same is true of persons who have been exposed to large amounts of radiation for medical reasons.

Gebhart (1970) has compared the effects of various drugs, most of them used in cancer chemotherapy, on cells of treated persons and on lymphocytes exposed in vitro.

However, most of the information we have about the induction of chromosome aberrations has been obtained from studies on cultured cells, both lymphocytes and fibroblasts. Cultured human cancer cells, especially various strains of HeLa cells (long-term human cell lines originating in a cervical carcinoma), have also been used in mutagenesis experiments. Numerous studies on the effects of chromosome-breaking agents have been performed on transplantable ascites tumors of the mouse, which are grown in a suspension in the abdominal cavity of the animal (cf. Adler, 1970). This method makes it easy to obtain beautiful chromosome preparations. Different types of mutagenicity tests on whole mammals have been reviewed by Russell and Matter (1980).

Various methods have been employed to determine chromosome damage (cf. Schoeller and Wolf, 1970; Nichols, 1973). One of the earlier techniques consisted of scoring dicentric bridges and acentric fragments in anaphase. Another method involved counting micronuclei formed by damaged or lagging chromosomes in mitoses following the treatment (cf. Nichols, 1973).

The most accurate results are naturally achieved by analyzing whole metaphase plates, especially by the various banding techniques. Even when this is done, however, the results of comparable studies conducted by different groups often yield discordant results. Apart from variation in the biological material and in culture techniques, the discrepancies depend mainly on different criteria used in scoring chromosome and chromatid aberrations. For example, one frequent source of confusion is the scoring of so-called gaps. These are mostly despiralized chromosome regions that appear as thinner and less well-stained regions in the chromatids. Since these gaps are difficult to delimit in either direction and as a rule do not lead to permanent chromosome damage, it would be best not to score them at all. Furthermore, since even a broken chromatid fragment tends to stick to its sister chromatid, it is often impossible to decide whether or not a true break has taken place. The dislocation of a fragment has commonly been used as a criterion to distinguish breaks from gaps (Schoeller and Wolf, 1970). Nevertheless, many scored breaks may, in reality, represent gaps. In addition, in most studies no definition or illustration is furnished to show the criteria according to which the various aberrations are classified.

In view of these difficulties, the most accurate procedure would be to score only dicentrics and quadriradials, since there can hardly be any disagreement about them. However, since these configurations require two breaks, they are much rarer than one-break aberrations. Consequently, to collect an adequate number often requires that a prohibitively large number of cells be checked or that the concentration of a drug or the amount of radiation used be so high as to interfere with cell division or cause cell death. Consequently, many recent studies on chromosome-breaking mutagens take advantage of the more sensitive system of sister chromatid exchanges, which is free from many of the difficulties just discussed. However, it should be kept in mind that the correlation of chromosome breaks and sister chromatid exchanges is far from perfect.

Rules for Chromosome-Breakage Studies

Increasing awareness of the dangers of radiation and of environmental poisons has given new impetus to studies on chromosome breakage. However, the results have often been controversial. Surprisingly, the results of studies on sister chromatid exchanges (Chapter X) have been equally inconsistent, although these exchanges should be easier to score than chromatid and chromosome breaks. At least some of the controversies might be resolved if certain rules were followed:

1) If the analysis is done on cultured cells, the control cells should be from the same individual, cultured at the same time. If this is not possible (for instance when whole-body irradiation is studied), the control person should be of the same age and sex. In animal studies, the controls should belong to the same inbred strain.
2) An illustration should show what is scored as a gap or a break, and preferably the other scored abnormalities should also be illustrated.
3) The chromosome slides of the treated and control cells, made and stained at the same time, ought to be randomly coded to avoid the bias of the investigator.
4) Clear rules should be established as to which metaphases are included in the study (how many chromosomes may be missing or extra). Furthermore, it should be stated whether the whole chromosome complement has been analyzed in all cells or only in those with obvious abnormalities.
5) The analysis should be done by a competent cytogeneticist (and not by part-time, inexperienced laboratory helpers).

References

Adler I-D (1970) Cytogenetic analysis of ascites tumour cells of mice in mutation research. In: Vogel F, Röhrborn G (eds) Chemical mutagenesis in mammals and man. Springer, New York, pp 251–259
Auerbach C (1976) Mutation research. Problems, results and perspectives.Chapman and Hall, London
Auerbach C (1978) Forty years of mutation research: a pilgrim's progress. Heredity 40: 177–187
Awa AA (1974) Cytogenetic and oncogenic effects of the ionizing radiations of the atomic bombs. In: German J (ed) Chromosomes and cancer. Wiley, New York, pp 637–674
Bijlsma JB, deFrance HF, Bleeker-Wagenmakers LM, et al (1978) Double translocation t(7;12),t(2;6) heterozygosity in one family. A contribution to the trisomy 12p syndrome. Hum Genet 40: 135–137
Bochkov NP, Kuleshov NP (1972) Age sensitivity of human chromosomes to alkylating agents. Mutat Res 14: 345–353
Coleman LC (1947) Chromosome abnormalities in an individual of *Chorthippus longicornis* (Acrididae). Genetics 32: 435–447
Evans HJ (1974) Effects of ionizing radiation on mammalian chromosomes. In: German J (ed) Chromosomes and cancer. Wiley, New York, pp 191–237
Gebhart E (1970) The treatment of human chromosomes in vitro: results. In: Vogel F, Röhrborn G (eds) Chemical mutagenesis in mammals and man. Springer, New York, pp 367–382
Harnden DG (1974) Viruses, chromosomes, and tumors: the interaction between viruses and chromosomes. In: German J (ed) Chromosomes and cancer. Wiley, New York, pp 151–190

Hayflick L, Moorhead PS (1961) The serial cultivation of human diploid cell strains. Exp Cell Res 25: 585-621

Kihlman BA (1966) Actions of chemicals on dividing cells. Prentice-Hall,Englewood Cliffs, New Jersey

Koskull H von, Aula P (1977) Distribution of chromosome breaks in measles, Fanconi's anemia and controls. Hereditas 87: 1-10

Kuhn EM, Therman E (1979) No increased chromosome breakage in three Bloom's syndrome heterozygotes. J Med Genet 16: 219-222

Nichols WW (1970) Virus-induced chromosome abnormalities. Annu Rev Microbiol 24: 479-500

Nichols WW (1973) Cytogenetic techniques in mutagenicity testing. Agents and actions 3: 86-92

Nichols WW (1975) Somatic mutation in biologic research. Hereditas 81: 225-236

Nichols WW (1983) Viral interactions with the mammalian genome relevant to neoplasia. In: German J (ed) Chromosome mutation and neoplasia. Liss, New York, pp 317-332

Rapp M, Therman E (1977) The effect of procarbazine on the chromosomes of normal and malignant mouse cells. Ann Génét 20: 249-254

Rieger R, Michaelis A (1967) Die Chromosomenmutationen. Gustav Fischer, Jena

Rieger R, Michaelis A (1972) Effects of chromosome repatterning in Vicia faba L. I. Aberration distribution, aberration spectrum, and karyotype sensitivity after treatment with ethanol of differently reconstructed chromosome complements. Biol Zbl 91: 151-169

Rieger R, Michaelis A, Nicoloff H (1982) Inducible repair processes in plant root tip meristems? "Below-additivity effects" of unequally fractionated clastogen concentrations. Biol Zbl 101: 125-138

Rieger R, Nicoloff H, Michaelis A (1973) Intrachromosomal clustering of chromatid aberrations induced by N-methyl-N-nitrosourethan in Vicia faba and barley. Biol Zbl 92: 681-689

Russell LB, Matter BE (1980) Whole-mammal mutagenicity tests: Evaluation of five methods. Mutat Res 75: 279-302

Schoeller L, Wolf U (1970) Possibilities and limitations of chromosome treatment in vitro for the problem of chemical mutagenesis. In: Vogel F, Röhrborn G (eds) Chemical mutagenesis in mammals and man. Springer, New York, pp 232-240

Schroeder TM, German J (1974) Bloom's syndrome and Fanconi's anemia: demonstration of two distinctive patterns of chromosome disruption and rearrangement. Humangenetik 25: 299-306

Schubert I, Michaelis A, Rieger R (1981) Effects of chromosome repatterning in Vicia faba L. V. Influence of segment transpositions on maleic hydrazide-specific aberration clustering on a heterochromatin containing chromosome region. Biol Zbl 100: 167-179

Schubert I, Rieger R (1980) Cytochemical and cytogenetic features of the nucleolus organizing region (NOR) of Vicia faba. Biol Zbl 99: 65-72

Seabright M, Gregson N, Pacifico E, et al (1978) Rearrangements involving four chromosomes in a child with congenital abnormalities. Cytogenet Cell Genet 20: 150-154

Shaw DD, Wilkinson P, Coates DJ (1983) Increased chromosomal mutation rate after hybridization between two subspecies of grasshoppers. Science 220: 1165-1167

Shaw MW, Cohen MM (1965) Chromosome exchanges in human leukocytes induced by mitomycin C. Genetics 51: 181-190

Therman E (1972) Chromosome breakage by 1-methyl–2-benzylhydrazine in mouse cancer cells. Cancer Res 32: 1133–1136

Therman E, Kuhn EM (1976) Cytological demonstration of mitotic crossing- over in man. Cytogenet Cell Genet 17: 254–267

White MJD (1963) Cytogenetics of the grasshopper *Moraba scurra* VIII. A complex spontaneous translocation. Chromosoma 14: 140–145

IX
Chromosome Breakage Syndromes

Genotypic Chromosome Breakage

In humans, chromosome aberrations, including nondisjunction and structural changes, increase with age. Another factor determining the frequency of chromosome aberrations is the genotype of the individual, some persons showing considerably higher rates than others.

Several genes are known that greatly increase the incidence of chromosome aberrations. The most extensively studied of such conditions, each caused by a different recessive autosomal gene, are Bloom's syndrome (BS), Fanconi's anemia (FA), and ataxia telangiectasia (AT), which has also been called Louis-Bar syndrome. In addition, in a few other diseases, claims of increased chromosome aberrations have been raised.

The great interest in BS, FA, and AT is out of proportion to their rare occurrence. The studies have been inspired by the hope that they would throw light on chromosome structure and behavior. Furthermore, the risk of cancer at an early age is greatly enhanced in such individuals: 1/4 of the BS patients and 1/8 of the AT patients have developed cancer, and in FA the risk is also estimated as 10–15 percent (cf. German, 1983). In none of these diseases is the biochemical or molecular basis known.

Bloom's Syndrome

The most typical features in BS are low birth weight and stunted growth. The patients often seek medical help, because they develop sun-sensitive telangiectatic erythema (dilation of blood vessels) (Fig. IX.1) (cf. Passarge, 1983). Bloom's Syndrome Registry now contains 103 patients of whom 25 have developed 28 cancers (German et al, 1984). The patients suffer from immunodeficiency, which often leads to respiratory tract

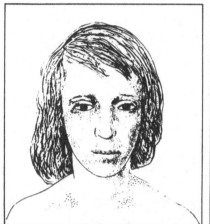
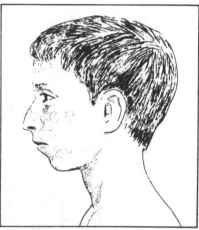

Figure IX.1. Characteristic face of a girl (left) and a boy (right) with Bloom's syndrome. Reproduced with permission from German, J. Oncogenic Implications of Chromosomal Instability. *Hospital Practice,* 8:2, 1973, p. 99.

infections. About half of the BS patients are of Jewish origin, and in two-thirds of the non-Jewish families, the parents are consanguineous.

German et al (1965) described a high frequency of chromosome aberrations in BS patients, of which most, possibly all, take place in S-G$_2$ (Kuhn and Therman, 1979; Therman and Kuhn, 1985). The characteristic feature, an enhanced tendency to exchanges between homologous chromosome segments, expresses itself in a 12-fold increase of sister chromatid exchanges (SCE) (Chaganti et al, 1974) (Fig. IX.2) and a 50- to 100-times increased incidence of mitotic crossing-over (cf. Therman and Kuhn, 1981).

A further interesting feature is that BS cells fuse spontaneously (Fig. XI.3); elsewhere this has been observed only when one cell type is malignant (Otto and Therman, 1982).

BS and normal cells have been grown together, or fused, to find out whether chromosome aberrations would decrease in BS cells or increase in normal cells. However, the results have been so inconsistent that no conclusion is possible at present.

Fanconi's Anemia

Known FA patients number over 300. They show a variety of symptoms, such as skeletal anomalies (often the thumbs are affected), hyperpigmentation of the skin, strabismus, and hypogonadism (cf. Alter and Potter, 1983). The patients develop lethal pancytopenia (lack of all blood elements) at the mean age of 5 years in boys and 6 years in girls.

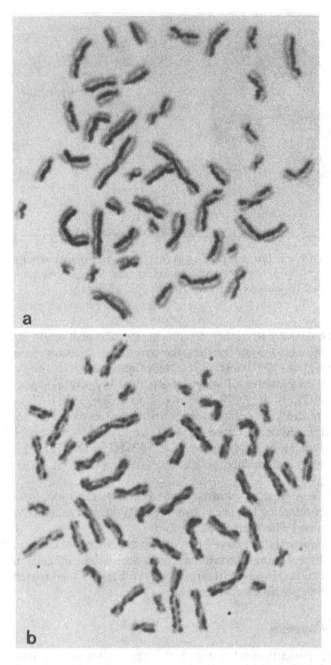

Figure IX.2. (a) Sister chromatid exchanges in a normal human lymphocyte; (b) highly increased rate of sister chromatid exchanges in a lymphocyte from a Bloom's syndrome patient (courtesy of RSK Chaganti). (See Chapter X for staining procedures.)

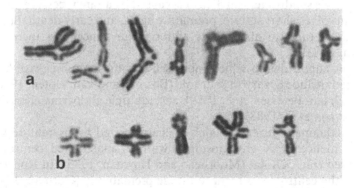

Figure IX.3. Characteristic chromosome aberrations in lymphocytes of a Fanconi's anemia patient. (a) Adjacent quadriradials; (b) alternate quadriradials (courtesy of EM Kuhn).

The most characteristic chromosome anomalies in FA are translocations between nonhomologous chromosomes (Fig. IX.3), which were first described by Schroeder et al (1964). Although the proportion of lymphocytes with aberrations may be as high as 30 percent, no increase in SCE has been observed (Chaganti et al, 1974).

Ataxia Telangiectasia

The most important symptoms in AT are progressive cerebellar ataxia, telangiectasia of eyes and skin, and severe immunodeficiency. Some 300 AT patients have now been described. Ataxia develops between 3 and 6 years of age, and the patients usually die of pulmonary infections or cancer. AT patients have a greatly increased sensitivity to x-rays (cf. Gatti and Hall, 1983).

Chromosome aberrations are less frequent in AT than in BS or FA. Originally random breakage apparently often leads to cell clones with a translocation, in most cases involving chromosome 14 (Kaiser McCaw et al, 1975). A characteristic translocation is t(14;14) (q12;q32) (cf. Kaiser-McCaw and Hecht, 1983). SCE are not increased.

Various types of lymphomas are the most common cancers in AT; in them the constant chromosome breakpoint is 14q32 (Chapter XXVI).

Other Conditions with Increased Chromosome Aberrations

Apart from BS, FA, and AT, increased chromosome aberrations have been described in some other diseases. However, in many cases adequate illustrations are lacking, and these claims should be reinvestigated in more detail.

Werner's syndrome, which also is caused by a rare recessive autosomal gene, involves short stature, premature aging, and early death. Both chromosome structural aberrations and risk for cancer are increased (cf. Brown, 1983).

Other conditions in which increased chromosome breakage has been reported include Sézary's syndrome (Bosman and van Vloten, 1976), psoriasis (Bruun Petersen et al, 1979), and multiple endocrine adenomatosis (Gustavson et al, 1983).

In Alzheimer's disease, which is characterized by premature senility, the incidence of X chromosomes with an inactivated centromere is increased (Fig. XX.4a) (Moorhead and Heyman, 1983). In Roberts' syndrome the centromeres seem to divide prematurely (German, 1979).

In xeroderma pigmentosum the chromosomes do not break spontaneously, but the excision repair of chromosome lesions caused by UV light is defective (cf. Pawsey et al, 1979). Exposure to light causes multiple skin lesions, which can develop into maligant neoplasms.

Cancer and Chromosome-Breakage Syndromes

The idea that chromosome structural aberrations cause cancer is further borne out by the chromosome breakage syndromes. In patients with these syndromes, the risk of malignant disease is greatly increased and cancer develops at a significantly earlier age than in the general population. In BS a further factor resulting from segregation of a mitotic chiasma is probably the high degree of homozygosity (Chapter X). The severe immunodeficiency in AT and the milder defect in BS are probably also cancer-promoting factors, since immunosuppressive drugs after organ transplants induce high rates of malignant neoplasms (cf. German, 1983).

References

Alter BP, Potter NU (1983) Long-term outcome in Fanconi's anemia: Description of 26 cases and review of the literature. In: German J (ed) Chromosome mutation and neoplasia. Liss, New York, pp 43–62

Bosman FT, van Vloten WA (1976) Sézary's syndrome: A cytogenetic, cytophotometric and autoradiographic study. J Pathol 118: 49–57

Brown WT (1983) Werner's syndrome. In: German J (ed) Chromosome mutation and neoplasia. Liss, New York, pp 85–93

Bruun Petersen G, Christiansen JV, Voetmann E, et al (1979) Chromosome aberrations in affected and unaffected skin of patients with psoriasis. Acta Dermatovener (Stockholm) 54: 147–151

Chaganti RSK, Schonberg S, German J (1974) A manyfold increase in sister chromatid exchanges in Bloom's syndrome lymphocytes. Proc Natl Acad Sci USA 71: 4508–4512

Gatti RA, Hall K (1983) Ataxia-telangiectasia: Search for a central hypothesis. In: German J (ed) Chromosome mutation and neoplasia. Liss, New York, pp 23–41

German J (1973) Implications of chromosomal instability. Hosp Pract 8: 93–104

German J (1979) Roberts' syndrome. I. Cytological evidence for a disturbance in chromatid pairing. Clin Genet 16: 441–447

German J (1983) Patterns of neoplasia associated with the chromosome-breakage syndromes. In: German J (ed) Chromosome mutation and neoplasia. Liss, New York, pp 97–134

German J, Archibald R, Bloom D (1965) Chromosomal breakage in a rare and probably genetically determined syndrome of man. Science 148: 506–507

German J, Bloom D, Passarge E (1984) Bloom's syndrome. XI. Progress report for 1983. Clin Genet 25: 166–174

Gustavson K-H, Jansson R, Öberg K (1983) Chromosomal breakage in multiple endocrine adenomatosis (types I and II). Clin Genet 23: 143–149

Kaiser-McCaw B, Hecht F (1983) The interrelationships in ataxia-telangiectasia of immune deficiency, chromosome instability, and cancer. In: German J (ed) Chromosome mutation and neoplasia. Liss, New York, pp 193–202

Kaiser McCaw B, Hecht F, Harnden DG, et al (1975) Somatic rearrangement of chromosome 14 in human lymphocytes. Proc Natl Acad Sci USA 72: 2071–2075

Kuhn EM, Therman E (1979) Chromosome breakage and rejoining of sister chromatids in Bloom's syndrome. Chromosoma 73: 275–286

Moorhead PS, Heyman A (1983) Chromosome studies of patients with Alzheimer disease. Am J Med Genet 14: 545–556

Otto PG, Therman E (1982) Spontaneous cell fusion and PCC formation in Bloom's syndrome. Chromosoma 85: 143–148

Passarge E (1983) Bloom's syndrome. In: German J (ed) Chromosome mutation and neoplasia. Liss, New York, pp 11–21

Pawsey SA, Magnus IA, Ramsay CA, et al (1979) Clinical, genetic and DNA repair studies on a consecutive series of patients with xeroderma pigmentosum. Q J Med, New Series 48, 190: 179–210

Schroeder TM, Anschütz F, Knopp A (1964) Spontane Chromosomenaberrationen bei familiärer Panmyelopathie.Humangenetik 1: 194–196

Therman E, Kuhn EM (1981) Mitotic crossing-over and segregation in man. Hum Genet 59: 93–100

Therman E, Kuhn EM (1985) Incidence and origin of symmetric and asymmetric dicentrics in Bloom's syndrome. Cancer Genet Cytogenet 15: 293–301

X

Sister Chromatid Exchanges and Mitotic Crossing-Over

The Detection of Sister Chromatid Exchanges

Sister chromatid exchanges (SCE) were discovered in the late 1950s with the same type of experiment as that which demonstrated the semiconservative replication of chromosomes (Chapter III). When cells are grown for one cycle in medium containing ^{3}H-thymidine and for another without it, autoradiography shows that one chromatid is labeled and the other is not (Figs. III.2, III.3); this makes exchanges between them visible (Taylor, 1958).

A more accurate technique is based on the observation that chromosomes in which thymidine is replaced by bromodeoxyuridine (BrdU) can be stained differentially (Fig. X.1) (Zakharov and Egolina, 1972). Three types of chromatids can be distinguished on the basis of whether both DNA strands, one, or none contains BrdU (Latt, 1974). Differential staining after various pretreatments can be done with Giemsa or with the fluorescent dyes Hoechst 33825, acridine orange, or coriphosphine O (Korenberg and Freedlender, 1974).

The Occurrence of SCE

Since both ^{3}H-thymidine and BrdU not only make SCE visible, but also induce them, the determination of the spontaneous exchange rate is not simple. That SCE do occur spontaneously is demonstrated by the behavior of ring chromosomes. Single rings often change into double dicentric rings through an uneven number of SCE (cf. Wolff, 1977).

In other types of cells the spontaneous SCE rate can only be estimated.

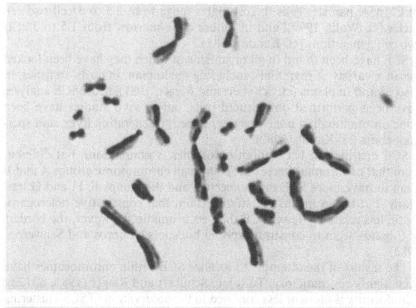

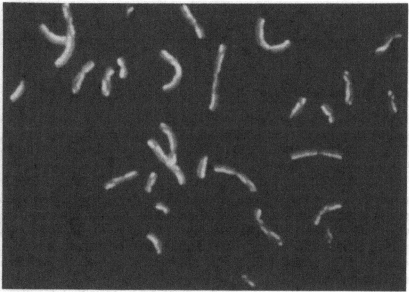

Figure X.1. Sister chromatid exchanges demonstrated with BrdU. Top: Giemsa-stained Chinese hamster cell (courtesy of JR Korenberg). Bottom: human lymphocyte stained with coriphosphine O.

In Chinese hamster cells the estimates range from 2.5 to 5/cell/two cell cycles (cf. Wolff, 1977), and in mouse bone marrow from 1.5 to 2/cell/two cell generations (cf. Kanda, 1982).

SCE have been found in all organisms in which they have been looked for: in a variety of mammals, including the human, in birds, in fishes, in insects, and in plants (cf. Schubert and Rieger, 1981). Most SCE analyses have been performed on cultured cells, but in vivo studies have been done on mammalian bone marrow, spleen, regenerating liver, and spermatogonia (cf. Kanda, 1982).

SCE distribution between chromosomes is nonrandom, but different from that of chromatid breaks. The human chromosome groups A and B seem to have more SCE than expected, and the groups E, F, and G less. Many, but by no means all, studies claim that constitutive heterochromatin has relatively fewer SCE than euchromatin. However, the borders of C-bands seem to constitute special hot spots (Ambros and Schweizer, 1983).

The results of the attempts to localize SCE within chromosomes have been highly contradictory. To quote Schubert and Rieger (1981, p. 122): "Conformity is more or less confined to the observation of SCE clustering in weakly fluorescent or Giemsa-stained bands and their borders." To summarize such results is almost impossible.

Clearly more observations with adequate statistical treatment are needed.

SCE in Mutagenesis Research

A wide variety of agents that, as a rule, also cause chromosome breaks have been found to induce SCE (cf. Latt, 1981; Schubert and Rieger, 1981; Sandberg, 1982). SCE have therefore become a major tool in mutagenesis research, as shown by the veritable flood of articles in this field.

The main advantage of SCE, compared with chromosome breaks, is that they are much more numerous and easier to score accurately. SCE are also more sensitive indicators of mutagenic activity. However, the two effects, chromosome breakage and SCE induction, do not always agree (cf. Schubert and Rieger, 1981). Furthermore, similar to many other aspects of SCE research, the results of mutagenesis studies have often been contradictory.

The highest incidence of SCE has been achieved with bifunctional alkylating agents. The most effective SCE inducers include methylmethanesulfonate, mitomycin C, dimethylsulfate, ethylmethanesulfonate, and the mustards (cf. Latt, 1981).

Ultraviolet light is also a powerful inducer of SCE. Of the chromosome breakage syndromes, only in BS are the SCE also increased (Fig. IX.2). Although AT and, even more, FA show a high incidence of chromosome aberrations (Chapter IX), no increase in SCE rates has been observed in them.

Similarly, ionizing radiation breaks chromosomes effectively, but increases SCE only slightly. A dose of x-rays that increases chromosome breaks 20-fold only doubles the SCE rate (cf. Kato, 1977).

An increase in SCE rate is certainly an indication of the mutagenic and carcinogenic potential of an agent. However, in view of the discrepancies between the induction of SCE and chromosome breaks, as well as of the many controversies in SCE research in general, a discovery of an increased SCE rate should be followed up by chromosome studies.

Significance of SCE

The basic events underlying SCE are unknown (cf. Latt, 1981). Obviously the mechanism is different from that causing chromatid breaks (cf. Schubert and Rieger, 1981). More probably SCE are related to meiotic and mitotic crossing-over. An important factor in all these exchanges seems to be the closeness of pairing between the homologs. In meiosis, where the pairing is most intimate, chiasma frequencies range from one to more than ten per bivalent. Next come the SCE whose rate is estimated as 1.5–5/cell/two cell generations. Diplochromosomes show 100 chiasmata/1000 cells, whereas mitotic chiasmata between two homologs, which probably pair only by accident, number 0.1–1/1000 cells (Therman et al, 1978). In BS this general tendency to exchanges is greatly increased, affecting both SCE and mitotic chiasmata.

The biological significance of SCE is unclear, since the genetic makeup of a cell is not changed, if the exchanges take place at identical points. However, *unequal* SCE lead to the amplification of genes on one chromatid and their deletion on the other. For instance, the variation of the heterochromatic segment on the Y chromosome obviously has come about through unequal SCE. The importance of gene duplication is discussed at the end of this chapter.

Mitotic Crossing-Over

Although there is little genetic evidence of mitotic crossing-over in mammals, including the human (for instance, blood group mosaicisms, twin color spots in fur or skin), a variety of cytological observations demonstrate that it takes place in human cells, especially in BS patients (German, 1964).

A mitotic chiasma in metaphase corresponds to a one-chiasma bivalent in meiotic metaphase (Fig. X.2A–I) or sometimes to a diplotene bivalent (Fig. X.2J). As in meiosis, the two homologs can also be joined by a terminal chiasma (Therman and Kuhn, 1976). That a mitotic chiasma represents true recombination is shown by heteromorphic "bivalents" in which the sister chromatids are attached to different centromeres (Patau and Therman, 1969; Therman and Kuhn, 1976); this has also been demonstrated by their pattern of SCE (Chaganti et al, 1974).

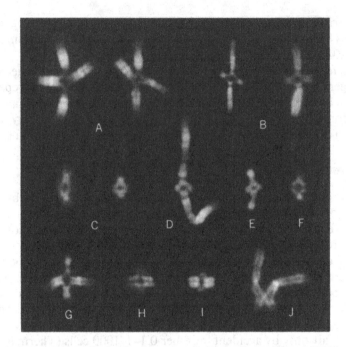

Figure X.2. Examples of mitotic chiasmata in Bloom's syndrome. (A) Two chiasmata in 1q; (B) two chiasmata in 6p; (C) two chiasmata in 19; (D) chiasma in 1q; (E) chiasma in 20q; (F) chiasma in 22q; (G) chiasma in 12q; (H) chiasma in 18q; (I) chiasma in centric region of 18; (J) chiasma in 3p (Kuhn, 1976).

That mitotic chiasmata are in principle different from chromatid trans-locations is demonstrated by a variety of observations (cf. Therman and Kuhn, 1981). Mitotic chiasmata are rare in normal human cells, ranging from 0.1 to 1/1000 cells, but are considerably more frequent in normal diplochromosomes. In BS lymphocytes they number 5–150/1000 cells, and the record is reached by BS diplochromosomes, which display 2500 chiasmata/1000 cells (Kuhn, 1981).

Adjacent configurations, which correspond to chiasmata, are of interest in that they are probably the main source for isodicentric chromosomes, including isochromosomes (Chapter VII) (Therman and Kuhn, 1985).

Nonrandom Localization of Chiasmata

The highly nonrandom distribution of mitotic chiasmata was demon-strated by Kuhn (1976), who analyzed 481 Q-banded chiasmata from BS. Most of them were situated in short Q-dark regions (or at their borders), special hot spots being 3p21, 6p21, 11q13, 12q13, 17q12, and 19p or q, and a high chiasma frequency was found also in the distal segments of 1p and 22q. About 17% of the chiasmata were situated in centric heterochromatin.

Segregation after Mitotic Recombination

The difference in segregation after meiotic and mitotic recombination is shown in Fig. X.3. In meiotic segregation the chromatids distal to the chiasma are nonhomologous (left), whereas after mitotic segregation they are sister chromatids (right), because human chromosomes show almost no relational coiling.

Visible demonstration of segregation after mitotic recombination is provided by distinct satellites. Exchanges between the repeated ribosomal RNA genes on the satellite stalks are especially frequent in BS (6/1000 cells). Thus, in a BS patient with Q-bright satellites, 31 different combinations were found in 58 cells, 12 cells being homozygous for the Q-bright satellites (Therman et al, 1981).

Obviously many other chromosome segments in BS have also become homozygous. This would allow presumed recessive cancer genes to express themselves and would explain the high incidence of cancer in BS (1/4 patients) (cf. Therman and Kuhn, 1981).

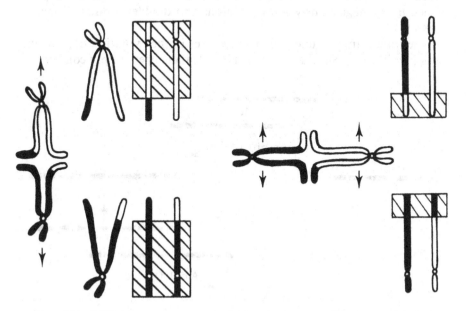

Figure X.3. Segregation after crossing-over in a meiotic bivalent (left) and a mitotic chiasma (right). In meiotic segregation the chromatids distal to the chiasma are nonsister, in mitotic segregation sister (Therman and Kuhn, 1981).

The Origin of Mitotic Chiasmata

Kuhn et al (1985) have shown that significantly more genes are localized to the chiasma-rich chromosomes than to similar-sized control chromosomes. Chiasma-rich chromosomes are also significantly rarer as trisomics in spontaneous abortions.The conclusion is that the chiasma-rich regions have a high density of active genes and are therefore extended in interphase (Korenberg et al, 1978). They would in consequence be more likely to pair and to cross over than other segments.

Gene Amplification

Unequal crossing-over, which may take place during meiosis, mitosis, or SCE, is an important mechanism in creating duplications and deficiencies (Fig. X.4). The prerequisite for unequal crossing-over is first an exchange between nonhomologous points, followed by mispairing of homologous segments for which repeated sequences have a special tendency.

The variation in heterochromatin within individuals or species, which variation probably is the result mainly of unequal crossing-over, has been assumed to be an important mechanism in speciation (cf. Flavell, 1982). The variation in the length of the satellite stalks is another result of unequal exchanges between repeated genes. A further example is provided by homogeneously stained regions and double minutes (Chapter XXV).

The importance of unequal crossing-over in the evolution of proteins was predicted by Smithies (1964). This idea has been amply confirmed.

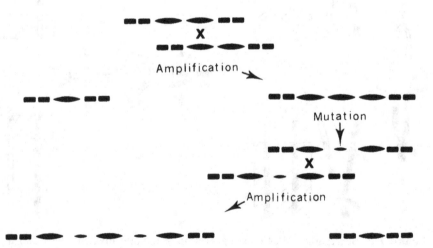

Figure X.4. Amplification and reduction of chromosome segments through unequal crossing-over.

Gene amplification has been found in a variety of organisms ranging from bacteria to human (cf. Schimke, 1982), and it is now clear that this process has played a major role in evolution.

References

Ambros P, Schweizer D (1983) Euchromatin-heterochromatin junctions are sister chromatid exchange (SCE) "hot spots". Kew chromosome conference II. George Allen and Unwin, London, p 326

Chaganti RSK, Schonberg S, German J (1974) A manyfold increase in sister chromatid exchanges in Bloom's syndrome lymphocytes. Proc Natl Acad Sci USA 71: 4508–4512

Flavell R (1982) Sequence amplification, deletion and rearrangement: Major sources of variation during species divergence. In: Dover GA, Flavell RB (eds) Genome evolution. Academic Press, New York, pp 301–323

German J (1964) Cytological evidence for crossing-over in vitro in human lymphoid cells. Science 144: 298–301

Kanda N (1982) Spontaneous sister chromatid exchange in vivo. In: Sandberg AA (ed) Sister chromatid exchange. Liss, New York, pp 279–296

Kato H (1977) Spontaneous and induced sister chromatid exchanges as revealed by the BUdR-labeling method. Int Rev Cytol 49: 55–97

Korenberg JR, Freedlender EF (1974) Giemsa technique for the detection of sister chromatid exchanges. Chromosoma 48: 355–360

Korenberg JR, Therman E, Denniston C (1978) Hot spots and functional organization of human chromosomes. Hum Genet 43: 13–22

Kuhn EM (1976) Localization by Q-banding of mitotic chiasmata in cases of Bloom's syndrome. Chromosoma 57: 1–11

Kuhn EM (1981) A high incidence of mitotic chiasmata in endoreduplicated Bloom's syndrome cells. Hum Genet 58: 417–421

Kuhn EM, Therman E, Denniston C (1985) Mitotic chiasmata, gene density, and oncogenes. Hum Genet 70: 1–5

Latt SA (1974) Localization of sister chromatid exchanges in human chromosomes. Science 185: 74–76

Latt SA (1981) Sister chromatid exchange formation. Annu Rev Genet 15: 11- 55

Patau K, Therman E (1969) Mitotic crossing-over in man. Genetics 61: Suppl 45–46

Sandberg AA (1982) Sister chromatid exchange in human states. In: Sandberg AA (ed) Sister chromatid exchange. Liss, New York, pp 619–651

Schimke RT (1982) Summary. In: Schimke RT (ed) Gene amplification. Cold Spring Harbor Laboratory, New York, pp 317–333

Schubert I, Rieger R (1981) Sister chromatid exchanges and heterochromatin. Hum Genet 57: 119–130

Smithies O (1964) Chromosomal rearrangements and protein structure. Cold Spring Harbor Symp Quant Biol 29: 309–319

Taylor JH (1958) Sister chromatid exchanges in tritium labeled chromosomes. Genetics 43: 515–529

Therman E, Denniston C, Sarto GE (1978) Mitotic chiasmata in human diplochromosomes. Hum Genet 45: 131–135

Therman E, Kuhn EM (1976) Cytological demonstration of mitotic crossing-over in man. Cytogenet Cell Genet 17: 254–267

Therman E, Kuhn EM (1981) Mitotic crossing-over and segregation in man. Hum Genet 59: 93–100

Therman E, Kuhn EM (1985) Incidence and origin of symmetric and asymmetric dicentrics in Bloom's syndrome. Cancer Genet Cytogenet 15: 293–301

Therman E, Otto PG, Shahidi NT (1981) Mitotic recombination and segregation of satellites in Bloom's syndrome. Chromosoma 82: 627–636

Wolff S (1977) Sister chromatid exchange. Annu Rev Genet 11: 183–201

Zakharov AF, Egolina NA (1972) Differential spiralization along mammalian mitotic chromosomes. I. BUdR-revealed differentiation in Chinese hamster chromosomes. Chromosoma 38: 341–365

XI

Cell Fusion, Prematurely Condensed Chromosomes, and the Origin of Allocyclic Chromosomes

Cell Fusion

Normal, untreated somatic cells do not, as a rule, fuse with one another. However, in exceptional cases this has happened between pollen mother cells in plant hybrids and between endosperm cells (only plant cells devoid of cell walls are able to fuse) (cf. Sperling and Rao, 1974). Rare spontaneous fusion between malignant mammalian cells was first described in the early 1960s (cf. Otto and Therman, 1982). Fusion of human cancer cells (Atkin, 1979) and leukemic cells (Knuutila et al, 1981) in vivo has also been observed. The only example of spontaneous fusion of nonmalignant human cells has so far been observed in lymphocytes and fibroblasts of Bloom's syndrome patients (Otto and Therman, 1982).

Infection by a variety of viruses causes cell fusion, which has led to the technique of cell fusion by treatment with inactivated Sendai virus (cf. Creagan and Ruddle, 1977). Nowadays, cell fusion is accomplished mainly with polyethylene glycol (PEG).

Prematurely Condensed Chromosomes (PCC)

The phenomenon that has attracted attention to cell fusion is premature chromosome condensation, which is also used to determine that cell fusion, and not some other polyploidization mechanism, has given rise to a polyploid cell. When a cell whose chromosomes are in metaphase is fused with a cell in interphase, the nuclear membrane of the interphase nucleus dissolves and the chromosomes condense. The condensation

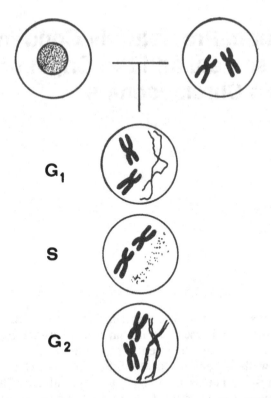

Figure XI.1. Fusion of an interphase and a metaphase cell and PCC formation. In G_1, the PCC are single and thin; in S they appear "pulverized"; in G_2 they resemble prophase chromosomes.

becomes visible some 15–20 min after the fusion and reaches its peak in 1 hour (cf. Rao, 1982; Sperling, 1982). The appearance of the PCC depends on the nuclear stage—G_1, S, or G_2—at the time of the fusion (Fig. XI.1). The PCC in early G_1 are short (Fig. XI.1), but become longer and thinner as G_1 progresses (Fig. XI.3). The chromosomes in G_0(interphase of cells that do not undergo mitosis) are similar to those in G_1.

Chromosomes condensed in the S phase have been called "pulverized" or "fragmented" because of their appearance. However, the chromosomes even at this stage are continuous, the gaps in them representing the segments that at the moment are in the process of replication. That the S type PCC, indeed, are synthesizing DNA can be demonstrated with ^{3}H-thymidine and autoradiography (Fig. XI.2). The G_2 type PCC resemble long, thin prophase chromosomes, which gradually shorten (Fig. XI.3, bottom). They can be banded with the usual techniques, but sometimes appear spontaneously banded (Fig. XI.3, bottom). The spontaneous

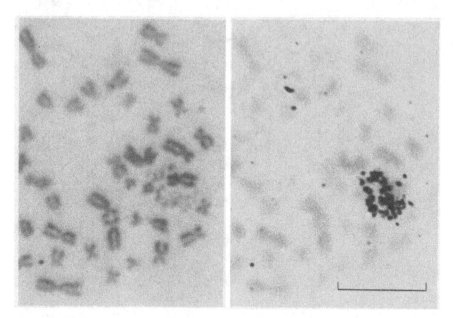

Figure XI.2. Part of a tetraploid lymphocyte metaphase from Bloom's syndrome with an allocyclic chromosome before (left) and after (right) autoradiography. ^{3}H-thymidine was added to the cell culture 4 hours before fixation (the plate also contains a mitotic chiasma) (Otto et al, 1981).

banding resembles that of chromosomes that have been fed BrdU in the late S period (Otto and Therman, 1982). Obviously, the late-replicating segments have not had enough time to coil and have remained stretched.

An interesting aspect of cell fusion and PCC formation is that cells from very different organisms can be induced to undergo these processes, the extremes being combinations like *Drosophila*-soybean and human-carrot cells (cf. Dudits et al, 1982). These experiments show that the same or similar factors are involved in chromosome condensation and mitosis in the widest variety of organisms. That this applies even to meiosis has been shown by experiments combining amphibian oocytes and human cells (Sunkara et al, 1982). The substances involved in all these processes are under intensive study.

Uses of PCC Formation

Cell fusion is a widely used tool in gene mapping (Chapter XXVIII) and in the creation of hybridomas. Cell fusion combined with PCC is an important technique in a variety of fields. PCC provide an opportunity to study interphase chromosomes and to follow the changes that they

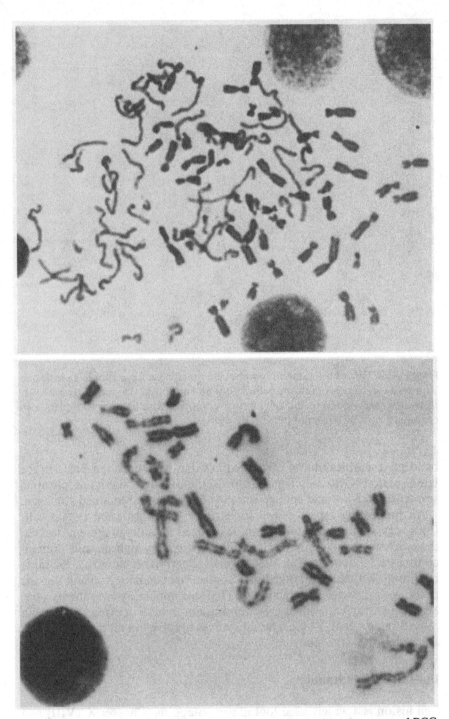

Figure XI.3. Spontaneous fusion of two Bloom's syndrome lymphocytes and PCC formation. Top: PCC in G_1; bottom: naturally banded PCC in G_2 (Otto and Therman, 1982).

undergo from G_1 through S and G_2. Since the chromosomes in late G_1 and early G_2 are longer than at any mitotic stage, their fine structure can be analyzed in greater detail. This also makes the determination of DNA and RNA synthesis more accurate (cf. Sperling, 1982).

An interesting aspect of the use of PCC is the study of differentiated cells, such as spermatids (Drwinga et al, 1979) or sperm, which do not divide spontaneously. These techniques might also be useful in the study of the cytogenetics of somatic cells. Furthermore, the arrangement of chromosomes in interphase nuclei can be followed by means of PCC (Chapter III). Such studies have confirmed that, in different types of mammalian cells, the chromosomes retain their polarized anaphase orientation (Rabl orientation) throughout interphase, but are otherwise situated at random (Sperling and Lüdtke, 1981; Cremer et al, 1982).

A further field in which PCC have been and probably will be increasingly useful is the study of chromosome breakage. With ordinary cytological techniques, chromosome breakage can be studied at the earliest several hours after its induction, whereas, with PCC, breaks can be analyzed in 20-30 min. This gives a good idea of the original breakage rate, before the majority of breaks have had time to be repaired. And, indeed, analysis of the PCC at different times after the treatment demonstrates a gradual decrease of visible chromosome aberrations (cf. Hittelman et al, 1980).

Origin of Allocyclic Chromosomes

An otherwise normal metaphase plate may, although extremely rarely, contain one or a few chromosomes that resemble PCC chromosomes in G_1, S, or G_2. The "pulverized" S type chromosomes are in the process of synthesizing DNA, while the rest of the chromosomes are in metaphase (Fig. XI.2). The rarity of such allocyclic chromosomes in normal, untreated cells is shown by the observation that in 2324 cells from 171 persons under 40 years of age who did not have a chromosome-breakage syndrome, one cell with an allocyclic chromosome was found (C Trunca, unpublished).

However, agents that break chromosomes also increase the incidence of allocyclic chromosomes (cf. Rao, 1977). In Bloom's syndrome, in which other chromosome abnormalities are common, the frequency of allocyclic chromosomes is greatly increased (Otto et al, 1981).

A commonly accepted hypothesis—in fact, no other explanation is usually considered—is that allocyclic chromosomes are derived from micronuclei in which the chromosomes condense prematurely under the influence of the main nucleus in metaphase (Kato and Sandberg, 1968; Obe and Beek, 1982). An alternative hypothesis has been proposed by Otto et al (1981). According to them, an allocyclic chromosome has undergone a mutation (possibly in a hypothetical coiling center), which renders it unable to keep up with the rest of the chromosomes. Such chromosomes have been left behind in their coiling cycle and will form micronuclei in the *next* mitosis.

The main argument for the micronucleus hypothesis is that the frequencies of allocyclic chromosomes and of micronuclei are similar. This would agree equally well with the idea that allocyclic chromosomes end up as micronuclei. However, further evidence for the micronucleus hypothesis is that the allocyclic chromosomes appear only at the second division after the induction of chromosome breakage (Obe and Beek, 1982).

On the other hand, several other observations, made mainly on Bloom's syndrome patients, are easily explained by the hypothesis of Otto et al (1981), but fit less well or not at all the micronuclear origin of allocyclic chromosomes: (1) In only 11/115 cells was the allocyclic chromosome lying at the rim of the metaphase plate. (2) In 24/115 metaphases, only a part of a chromosome was "pulverized". (3) A "pulverized" acrocentric was often found to be in satellite association with another acrocentric. (4) When a ring chromosome is replaced by an allocyclic chromosome, as often happens, it is practically always a member of a 46-chromosome complement and not an extra chromosome, as it should be equally frequently if it arose from a micronucleus. (5) Allocyclic chromosomes occur 5 times more frequently in cells with other chromosome abnormalities. (6) Allocyclic chromosomes are found in 16% of the tetraploid, but in only 2% of the diploid, cells.

On the basis of present evidence, it seems probable that allocyclic chromosomes represent both those chromosomes that have been left behind in their cycle, and micronuclei that have undergone PCC. Future observations should be analyzed in the light of both hypotheses to find out which is in better agreement in each case.

References

Atkin NB (1979) Premature chromosome condensation in carcinoma of the bladder: presumptive evidence for fusion of normal and malignant cells. Cytogenet Cell Genet 23: 217–219

Creagan RP, Ruddle FH (1977) New approaches to human gene mapping by somatic cell genetics. In: Yunis JJ (ed) Molecular structure of human chromosomes. Academic, New York, pp 89-142

Cremer T, Cremer C, Baumann H, et al (1982) Rabl's model of the interphase chromosome arrangement tested in Chinese hamster cells by premature chromosome condensation and laser-UV-microbeam experiments. Hum Genet 60: 46–56

Drwinga HL, Hsu TC, Pathak S (1979) Induction of prematurely condensed chromosomes from testicular cells of the mouse. Chromosoma 75: 45–50

Dudits D, Szabados L, Hadlaczky G (1982) Premature chromosome condensation in plant cells and its potential use in genetic manipulation. In: Rao PN, Johnson RT, Sperling K (eds) Premature chromosome condensation. Academic, New York, pp 359-369

Hittelman WN, Sognier MA, Cole A (1980) Direct measurement of chromosome damage and its repair by premature chromosome condensation. In: Meyn RE, Withers HR (eds) Radiation biology in cancer research. Raven, New York, pp 103-123

Kato H, Sandberg AA (1968) Chromosome pulverization in human cells with micronuclei. J Nat Cancer Inst 40: 165-179

Knuutila S, Siimes M, Vuopio P (1981) Chromosome pulverization in blood diseases. Hereditas 95: 15-24

Obe G, Beek B (1982) Premature chromosome condensation in micronuclei. In: Rao PN, Johnson RT, Sperling K (eds) Premature chromosome condensation. Academic, New York, pp 113-130

Otto PG, Otto PA, Therman E (1981) The behavior of allocyclic chromosomes in Bloom's syndrome. Chromosoma 84: 337-344

Otto PG, Therman E (1982) Spontaneous cell fusion and PCC formation in Bloom's syndrome. Chromosoma 85: 143-148

Rao PN (1977) Premature chromosome condensation and the fine structure of chromosomes. In: Yunis JJ (ed) Molecular structure of human chromosomes. Academic, New York, pp 205-231

Rao PN (1982) The phenomenon of premature chromosome condensation. In: Rao PN, Johnson RT, Sperling K (eds) Premature chromosome condensation. Academic, New York, pp 1-41

Sperling K (1982) Cell cycle and chromosome cycle: morphological and functional aspects. In: Rao PN, Johnson RT, Sperling K (eds) Premature chromosome condensation. Academic, New York, pp 43-78

Sperling K, Lüdtke E-K (1981) Arrangement of prematurely condensed chromosomes in cultured cells and lymphocytes of the Indian muntjac. Chromosoma 83: 541-553

Sperling K, Rao PN (1974) The phenomenon of premature chromosome condensation: its relevance to basic and applied research. Humangenetik 23: 235-258

Sunkara PS, Wright DA, Adlakha RC, et al (1982) Characterization of chromosome condensation factors of mammalian cells. In: Rao PN, Johnson RT, Sperling K (eds) Premature chromosome condensation. Academic, New York, pp 234-251

XII
Modifications of Mitosis

Although normal mitosis is characterized by the regular alternation of chromosome reproduction and segregation of daughter chromosomes, the two processes are not necessarily correlated, and their relationship can be changed in different ways (cf. Oksala, 1954). As a rule, such modifications lead to a chromosome constitution differing from the basic complement of the individual (Fig. XII.1). Apart from multipolar mitoses, all other mitotic modifications are characterized by an absent or defective spindle, and in most cases these result in the duplication of the chromosome number (Fig. XII.2). Nagl (1978) has united the most important of these mechanisms, endoreduplication and endomitosis, under the term endocycles. The terminology referring to mitotic modifications has been the subject of considerable dispute. I shall use that proposed by Levan and Hauschka (1953), since it is established in the literature.

Endoreduplication

The most common modification of mitosis is endoreduplication, in which the chromosomes replicate two or more times between two mitoses instead of once as in normal mitosis (cf. Levan and Hauschka, 1953). If two replications have occurred, the chromosomes in a subsequent mitosis consist of four chromatids instead of the usual two. As mentioned in an earlier chapter, such structures are called diplochromosomes (Fig. XXII.4). After three or four endoreduplications, bundles consisting of 8 or 16 chromatids, respectively, can be observed (Fig. XII.3c).

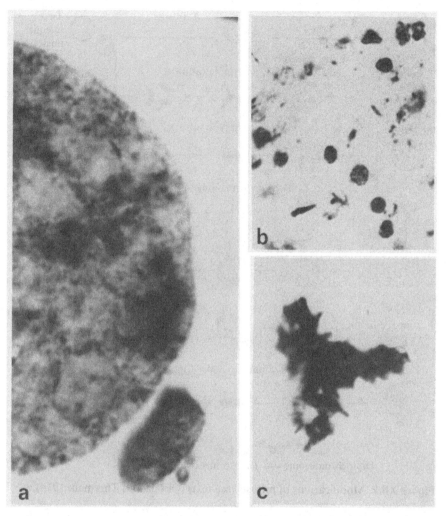

Figure XII.1. (a) Half of a giant nucleus from a human cervical cancer cell compared with (b) diploid stroma nuclei. (c) Side view of a tripolar metaphase from a mouse cancer cell.

In differentiated cells one endoreduplication after another may take place, so that the nucleus increases stepwise in both size and degree of polyploidy. Enormous nuclei, which obviously have come about through repeated endoreduplications, are characteristic especially of secretory cells and of tumor cells in both plants and animals (Fig. XII.1).

Endoreduplication probably occurs occasionally in all tissues. For instance, in human fibroblast cultures 3–5 percent of the dividing cells show a tetraploid chromosome number (a few are even octoploid),

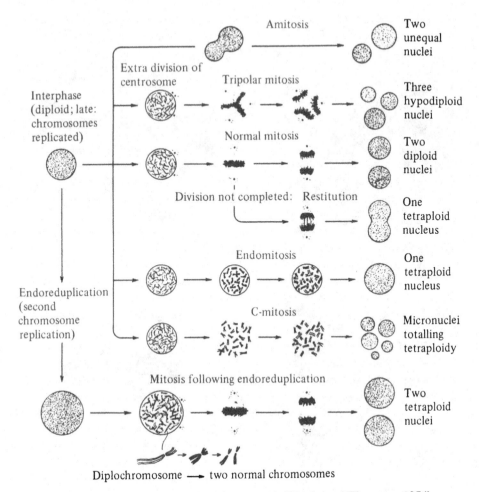

Figure XII.2. Modifications of mitosis (see text) (Oksala and Therman, 1974).

whereas such divisions are rare in cultured lymphocytes. Diplo-chromosomes are found in only a fraction of the tetraploid divisions, however, since they occur only in the first division after endoreduplica-tion. In differentiated tissues, on the other hand, endoreduplication is the most common mode of polyploidization.

Polyteny

Polytenization is a modification of endoreduplication in which the rep-licated homologous chromatids remain paired. The best known polytene chromosomes are the giant chromosomes of dipteran salivary glands (Fig. XII.4c).

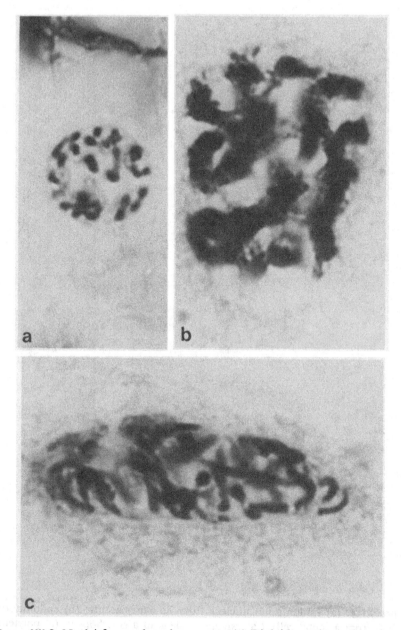

Figure XII.3. Nuclei from cultured pea roots. (a) Diploid prophase; (b) polytene chromosomes without banding; (c) nucleus with multiple chromatid bundles (Therman and Murashige, 1984).

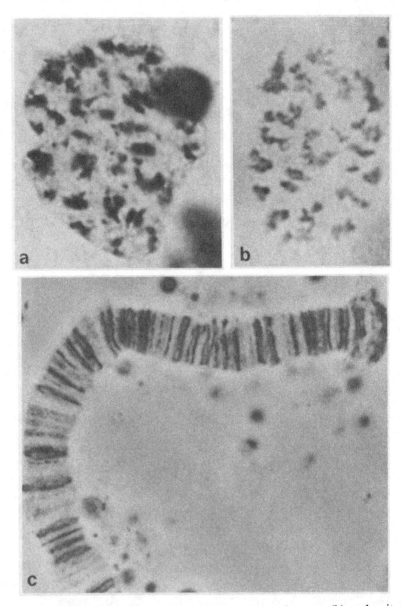

Figure XII.4. (a) Endomitotic nucleus from human placenta; (b) endomitotic nucleus from a septal cell of a testicular follicle in a grasshopper; (c) polytene chromosome from *Drosophila melanogaster* (c courtesy of R Kreber).

Modifications of these typically giant chromosomes are found in many other insect tissues (cf. Beerman, 1962). A polytene chromosome is composed of partially uncoiled chromosome strands paired side-by-side. At certain points, called chromomeres, the strand is folded. The chromomeres are aligned and form the visible bands in the polytene chromosome. In the diptera, each polytene chromosome consists of the two homologues. The stretched-out parts of the chromosome strands are the interband regions, whereas each band is thought to contain one gene. The chromosomes replicate time and again, and a dipteran salivary chromosome may finally consist of as many as 16,000 units (Beerman, 1962).

During the development of a tissue, certain bands specific for the tissue form so-called puffs. Puffing involves the unraveling of the coiled DNA of a particular chromomere; RNA is being synthesized at the puffs, and this demonstrates that the puffed genes are active in specific tissues during specific periods in development.

Polytene chromosomes are not limited to insects but are found in certain specialized plant cells, such as the synergids and antipods in the embryo sac as well as in suspensor and endosperm cells (cf. Tschermak-Woess, 1956; Nagl, 1981). Polytene chromosomes induced by culture conditions in pea roots (ordinarily pea roots consist of 2C and 4C cells) are illustrated in Fig. XII.3b (Therman and Murashige, 1984).

Interestingly, typical polytene chromosomes have recently been described in the giant trophoblast cells of the rabbit, mouse, and rat (Zybina et al, 1975). The banded structure is usually much less clear in non-dipteran polytene chromosomes, since mitotic chromosomes tend to pair only in diptera. However, good banding has been obtained in the bean *Phaseolus* by growing the plants at low temperatures (cf. Nagl, 1976).

Endomitosis

In *endomitosis*, prophase is normal. However, the nuclear membrane never dissolves, and the chromosomes continue to contract within it (Fig. XII.2). After they have reached their maximum contraction in endometaphase, the sister chromatids separate in endoanaphase but do not move far apart. The chromosomes then undergo telophasic changes and the nucleus reverts to interphase, having doubled its chromosome number.

This was the classic description of endomitosis by Geitler (1939) in various tissues of the heteropteran insect *Gerris*. However, it has recently become clear that other types of endomitosis exist. In the septal cells of the testicular follicles of grasshoppers, no DNA synthesis, but intensive RNA synthesis, takes place in the endomitotic cells (Fig. XII.4b) (Kiknadze and Istomina, 1980; Therman et al, 1983a). Such cells have obviously reached the final stage in their differentiation and are specialized for manufacturing some gene products.

In endomitotic cells of human hydatidiform moles, the chromosomes replicate nonsynchronously, part of them remaining condensed at all times. Thus a typical prophase and interphase are absent (Therman et al, 1985). Clearly endomitosis in various organisms should be restudied with modern techniques.

In the salivary gland of the insect *Gerris*, the cells become 1024-ploid to 2048-ploid as a result of successive endomitoses. Endomitosis is also found in the tissues of many other insects as well as of other animal groups (cf. Geitler, 1953; Heitz, 1953; Nagl, 1978). The tapetum cells of anther lobes in many plants undergo endomitosis (cf. D'Amato, 1952; Oksala and Therman, 1977). It has also been repeatedly described in cancer cells of both mouse and humans (Levan and Hauschka, 1953; Oksala and Therman, 1974; Therman et al, 1983b), but the only *normal* human tissue in which it has so far been observed is the placenta (Fig. XII.4a) (Sarto et al, 1982).

C-Mitosis

In C-mitosis (named after the drug colchicine) the chromosomes behave normally throughout prophase and up to metaphase. However, the spindle is defective or absent so that the chromosomes do not collect into a metaphase plate. Scattered around the cell, they divide and thereafter either unite into one tetraploid restitution nucleus or form a number of micronuclei of variable sizes (Fig. XII.2). Such micronuclei are usually unable to divide further. C-mitosis is extremely rare in normal cells, but it has been described often in mammalian cancer cells, where it probably reflects anoxia and tissue degeneration.

Restitution

Still another mitotic abnormality that leads to the doubling of the chromosome number is restitution. If the two groups of daughter chromosomes fail to separate in anaphase, for example, they may form one dumbbell-shaped nucleus (Fig. XII.2). Restitution may also take place in prophase or metaphase; or, if a cell has two nuclei that divide simultaneously (a common phenomenon in plant tapetum and mammalian liver cells), two metaphase or anaphase plates may fuse. Restitution may occur very occasionally in normal human cells, but it is not rare in malignant tumors.

Multipolar Mitoses

A mitotic aberration that can be regarded as the reverse of an absent or defective spindle is the formation of multipolar spindles. The most common are spindles with three poles (Figs. XII.1 and XII.2). The next in frequency are quadripolar mitoses, and spindles with larger numbers of poles are correspondingly rarer. Most tripolar and quadripolar spindles come about by one or both centrosomes dividing twice during a mitotic cycle, and not by cell and nuclear fusions, as assumed earlier (Therman and Timonen, 1950). Like so many other mitotic abnormalities, multipolar divisions are frequent in malignant tumors but practically nonexistent in untreated normal cells (cf. Oksala and Therman, 1974). If a multipolar anaphase is followed by cell division (often prevented by restitution), the resulting three or more daughter cells have abnormal chromosome constitutions and usually will not divide again.

References

Beerman W (1962) Riesenchromosomen. In: Protoplasmatologia, Handbuch der Protoplasmaforschung, Vol VI, D. Springer, Vienna

D'Amato F (1952) Polyploidy in the differentiation and function of tissues and cells in plants. Caryologia 4: 311–358

Geitler L (1939) Die Entstehung der polyploiden Somakerne der Heteropteren durch Chromosomenteilung ohne Kernteilung. Chromosoma 1: 1–22

Geitler L (1953) Endomitose und endomitotische Polyploidisierung. In: Protoplasmatologia, Handbuch der Protoplasmaforschung, Vol VI, C. Springer, Vienna

Heitz E (1953) Über intraindividuale Polyploidie. Arch Julius Klaus- Stiftung 28: 260–271

Kiknadze II, Istomina A G (1980) Endomitosis in grasshoppers. I. Nuclear morphology and synthesis of DNA and RNA in the endopolyploid cells of the inner parietal layer of the testicular follicle. Eur J Cell Biol 21: 122–133

Levan A, Hauschka S T (1953) Endomitotic reduplication mechanism in ascites tumors of the mouse. J Nat Cancer Inst 14: 1–46

Nagl W (1976) Nuclear organization. Annu Rev Plant Physiol 27: 39–69

Nagl W (1978) Endopolyploidy and polyteny in differentiation and evolution. Elsevier/North-Holland, Amsterdam

Nagl W (1981) Polytene chromosomes of plants. Int Rev Cytol 73: 21–53

Oksala T (1954) Timing relationships in mitosis and meiosis. Caryologia (Suppl) 6: 272–281

Oksala T, Therman E (1974) Mitotic abnormalities and cancer. In: German J (ed) Chromosomes and cancer. Wiley, New York, pp 239–263

Oksala T, Therman E (1977) Endomitosis in tapetal cells of Eremurus (Liliaceae). Am J Bot 64: 866–872

Sarto G E, Stubblefield P A, Therman E (1982) Endomitosis in human trophoblast. Hum Genet 62: 228–232

Therman E, Murashige T (1984) Polytene chromosomes in cultured pea roots (Pisum, Fabaceae). Pl Syst Evol 148: 25–33

Therman E, Sarto G E, Buchler D A (1983a) The structure and origin of giant nuclei in human cancer cells. Cancer Genet Cytogenet 9: 9–18

Therman E, Sarto G E, Kuhn E M (1985) The course of endomitosis in human cells. Cancer Genet Cytogenet (in press)

Therman E, Sarto G E, Stubblefield, P A (1983b) Endomitosis: a reappraisal. Human Genet 63: 13–18

Therman E, Timonen S (1950) Multipolar spindles in human cancer cells. Hereditas 36: 393–405

Tschermak-Woess E (1956) Karyologische Pflanzenanatomie. Protoplasma 46: 798–834

Zybina E V, Kudryavtseva M V, Kudryavtsev B N (1975) Polyploidization and endomitosis in giant cells of rabbit trophoblast. Cell Tiss Res 160: 525–537

XIII

Somatic Cell Cytogenetics

History of Somatic Cell Cytogenetics

By somatic cell cytogenetics is meant the study of differentiated cells outside the germline. Since such cells usually do not divide, it is not surprising that much less is known about them than about germline or meristematic cells.

Somatic cells began attracting the attention of cytologists after the realization that the giant polytene chromosomes in dipteran salivary glands were, indeed, chromosomes (Heitz and Bauer, 1933; Painter, 1933). During the 1930s and 1940s somatic cells were studied especially in plants and insects. Major credit for this work goes to the Viennese school under the direction of Geitler (cf. 1953) and Tschermak-Woess (cf. 1971). Recent reviews of the field are found in D'Amato (1977, 1984) and Nagl (1978).

Methods of Somatic Cell Cytogenetics

Only meristematic somatic cells divide regularly. Other cells that do not undergo mitosis spontaneously can be induced to do so by various means, such as wounding or treatment with different chemicals. For instance, the chromosomes in mouse liver, in whose development endopolyploidy seems to play an important role, can be studied by removing a part of the liver and fixing the cells when the tissue begins to regenerate. Mouse liver cells can also be induced to divide by injecting the animal with carbon tetrachloride (CCl_4). Differentiated plant cells also renew their mitotic activity around a wound or after treatment with plant-growth substances (cf. Therman, 1951; D'Amato, 1952; Tschermak-Woess, 1956).

However, highly polyploid cells usually cannot be induced to divide. Therefore, other techniques have to be used in their analysis. Microscopic examination combined with special staining techniques or autoradiography reveals the size and shape of nuclei, chromocenters and condensed chromatin, as well as endomitosis and polytene chromosomes. Electron microscopy has also been useful (Nagl, 1982b).

Measurements of nuclei have largely been replaced by spectrophotometric DNA determinations (Patau and Srinivasachar, 1960). The most recent technique for the determination of the DNA content of nuclei is flow cytometry (Chapter IV). Cell fusion and formation of prematurely condensed chromosomes may also have an important future in the study of differentiated cells (Chapter XI).

Somatic Polyploidy

The most common cytological difference between the germline and soma is that in most organisms somatic tissues consist of polyploid cells or are mosaics of diploid and of different types of polyploid cells. Species or larger taxa that lack endopolyploidy (for example, conifers and compositous plants) are exceptions (cf. Nagl, 1978).

Differentiated plant tissues are usually mosaics of diploid, tetraploid, and more highly polyploid cells. Consider an onion root. In the growing point, which occupies a few millimeters of the tip, the cells are diploid and divide normally. Above this meristematic region is a zone where cell divisions have stopped. The cells grow in size, and the nuclei attain various degrees of ploidy through endoreduplication. Above this region the differentiated cells, which do not change further, form a mosaic of diploid, tetraploid, octoploid, and even more highly polyploid cells (Therman, 1951). In certain plants, the best studied of which is spinach, cells still divide in the zone of differentiation, and the metaphases exhibit 2n, 4n, and 8n diplochromosomes in addition to diploid chromosome constitutions.

In insects the entire process of differentiation is correlated with endopolyploidy, each tissue representing more or less the same level of ploidy. Considerably less is known about mammalian tissues. However, endopolyploidy is characteristic at least of liver, placenta, megakaryocytes,and myocardium (cf. Nagl, 1978).

When a tissue to some extent escapes the forces of developmental controls, as is the case with ephemeral organs such as the anther tapetum in plants, the placenta in mammals, or abnormal tissues such as hydatidiform moles, the variety of cells and the mechanisms giving rise to them show a considerably wider range than is the case in normal tissues. This is even more true of malignant tumors, which seem to have escaped con-

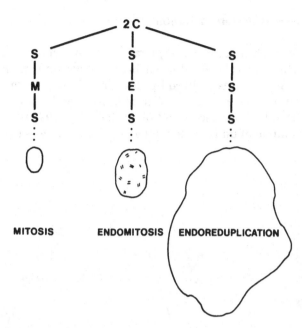

Figure XIII.1. Modes of growth in hydatidiform moles (S = DNA synthesis; M = mitosis; E = endomitosis) (Therman et al, 1985).

trolling mechanisms completely (cf. Oksala and Therman, 1974; Therman et al, 1983; Therman et al, 1984). Indeed, malignant tumors exemplify all possible mitotic aberrations and the resulting variable cell types.

Human hydatidiform moles (Chapter XXI) provide a good example of abnormal, but not malignant growth; the mechanisms include cell division, endomitosis, and endoreduplication (Fig. XIII.1) (Sarto et al, 1984; Therman et al, 1985). The same modes of growth are found in malignant tumors (Therman et al, 1984) and probably in many normal tissues.

Structure of Nuclei in Somatic Cells

The shapes of nuclei in somatic cells vary from spherical to elongated to lobed and even to branched structures. Another factor determining their appearance is the amount of condensed chromatin. In plants, condensed chromatin is species-specific (Nagl, 1982b). In mammalian cells, on the other hand, chromatin condensation varies greatly between different tissues, the pattern being tissue-specific (Nagl, 1982b). In some (female) cells, only the X chromosome is condensed; in others most of the chromosome complement appears to be in a condensed state (Fig. V.2). Obviously, in mammals, chromosome condensation provides a mechanism for the inactivation of those chromosome segments whose function is not needed in a particular cell type.

Amplification and Underreplication

That the heterochromatin in dipteran polytene chromosomes has remained at the diploid level, while euchromatin has undergone one replication after another, was realized by Heitz in 1933 (cf. Beerman, 1962). Since then, numerous examples of amplification and underreplication have been found both in insects and plants (cf. Nagl, 1982a).

The first example of underreplication of euchromatin was described by

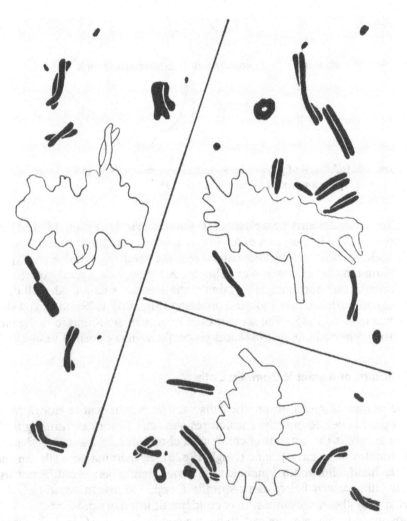

Figure XIII.2. Three polyploid metaphases in side view from differentiated onion root cells containing numerous broken chromosomes, acentric fragments, and ring chromosomes (Therman, 1951).

Therman (1951). When the differentiated cells in the onion roots are induced to divide with indol–3-acetic acid, the diploid mitoses show normal chromosomes, whereas in tetraploid and more higly polyploid cells, the chromosomes have undergone extensive breakage and rejoining (Fig. XIII.2). This was interpreted by Therman (1951) to mean that only those genes that are needed replicate, while the others remain at the diploid level. When such chromosomes are compelled to divide, they naturally break at the underreplicated regions.

Amplification may occur in tandem, leading to homogeneously stained regions (Chapter XXV) or "laterally"in differentiated cells. Apart from heterochromatin, the latter type of amplification often involves ribosomal RNA genes in oocytes of many organisms but also in other cell types (cf. Nagl, 1982a).

Somatic Cell Cytogenetics and Differentiation

That there is a correlation between endopolyploidy and differentiation is obvious. However, the exact relationship of these two phenomena is still a matter of speculation (cf. Nagl, 1981).

A polyploid cell is in many ways different from the corresponding diploid. This is demonstrated by unicellular organisms, such as the green alga *Micrasterias*, in which the duplication of the chromosome complement is reflected both in morphological and physiological changes in the cell (cf. Kallio, 1953). In addition to the change in the ratio of nuclear surface to volume, some genes may, in duplicate, have more pronounced effects than others. Polyploidy may further affect cell cycle duration and the size of cells and organs (cf. Nagl, 1981).

Beerman (1962) suggested that it was more economical for the organism to increase the number of genes through polyploidization than through the process of chromosome replication, mitosis, and cell division. The synchronization of gene activity is also probably easier in a polytene chromosome than in thousands of diploid cells. Furthermore, RNA synthesis can continue uninterrupted by mitosis (cf. Nagl, 1981). These ideas are in agreement with the observation that glandular and other active cells show especially high levels of polyploidy. The high multiplicity of dipteran salivary gland chromosomes was mentioned in Chapter VI. The trophoblast cells of many mammals are also highly polyploid. In the rat such cells reach a DNA content of 4,096C, the suspensor cells in plant embryos vary between 1,000C and 8,000C, and the record seems to be held by the silk gland cells of *Bombyx mori* at 524,288C corresponding to 19 duplications (cf. Nagl, 1982a).

A further advantage of endopolyploidy is that amplification and underreplication of chromosome regions, which would not be possible in mitotic cells, can be accomplished during polyploidization (cf. Nagl,

1982a). Although it is highly probable that the general structure of differentiated nuclei, such as polyteny, endomitosis, or various types of condensed chromatin, reflects different functions, so far little is known about the exact correlations of structure and function.

References

Beerman W (1962) Riesenchromosomen. In: Protoplasmatologia, Handbuch der Protoplasmaforschung, Vol VI, D. Springer, Vienna

D'Amato F (1952) Polyploidy in the differentiation and function of tissues and cells in plants. Caryologia 4: 311–358

D'Amato F (1977) Nuclear cytology in relation to development. University Press, Cambridge, England

D'Amato F (1984) Role of polyploidy in reproductive organs and tissues. In: Johri BM (ed) Embryology of angiosperms. Springer, Heidelberg, pp 519- 566

Geitler L (1953) Endomitose und endomitotische Polyploidisierung. In: Protoplasmatologia, Handbuch der Protoplasmaforschung, Vol VI, C. Springer, Vienna

Heitz E, Bauer H (1933) Beweise für die Chromosomennatur der Kernschleifen in den Knäuelkernen von *Bibio hortulanus* L. Z Zellforsch 17: 67–82

Kallio P (1953) On the morphogenetics of the desmids. Bull Torrey Bot Club 80: 247–263

Nagl W (1978) Endopolyploidy and polyteny in differentiation and evolution. Elsevier/North-Holland, Amsterdam

Nagl W (1981) Polytene chromosomes of plants. Int Rev Cytol 73: 21–53

Nagl W (1982a) Cell growth and nuclear DNA increase by endoreduplication and differential DNA replication. In: Nicolini C (ed) Cell growth. Plenum, New York, pp 619–651

Nagl W (1982b) Condensed chromatin:Species-specificity, tissue-specificity and cell cycle-specificity, as monitored by scanning cytometry. In: Nicolini C (ed) Cell growth. Plenum, New York, pp 171–218

Oksala T, Therman E (1974) Mitotic abnormalities and cancer. In: German J (ed) Chromosomes and cancer. Wiley, New York, pp 239–263

Painter T S (1933) A new method for the study of chromosome rearrangements and the plotting of chromosome maps. Science 78: 585–586

Patau K, Srinivasachar D (1960) A microspectrophotometer for measuring the DNA-content of nuclei by the two wave length method. Cytologia 25: 145–151

Sarto G E, Stubblefield P A, Lurain J, et al (1984) Mechanisms of growth in hydatidiform moles. Am J Obstet Gynecol 148: 1014–1023

Therman E (1951) The effect of indole–3-acetic acid on resting plant nuclei. I. *Allium cepa*. Ann Acad Sci Fenn A IV 16: 1–40

Therman E, Buchler D A, Nieminen U, et al (1984) Mitotic modifications and aberrations in human cervical cancer. Cancer Genet Cytogenet 11: 185–197

Therman E, Sarto G E, Kuhn E M (1985) The course of endomitosis in human cells. Cancer Genet Cytogenet (in press)

Therman E, Sarto G E, Stubblefield P A (1983) Endomitosis: a reappraisal. Hum Genet 63: 13–18

Tschermak-Woess E (1956) Karyologische Pflanzenanatomie. Protoplasma 46: 798–834

Tschermak-Woess E (1971) Endomitose. In: Handbuch der Allgemeinen Pathologie. Springer, Heidelberg, pp 569–625

XIV
Main Features of Meiosis

Significance of Meiosis

The most important modification of mitosis is meiosis, which is the reduction division that gives rise to the haploid generation in the life cycle. In mammals, the haploid generation is restricted to one cell type, the gamete, whereas the other cells in the animal are diploid. However, in many other organisms, especially lower plants, the haploid generation is more important than the diploid; or the two generations may be more or less equal, as in the mosses. In those organisms in which the haploid phase dominates, the diploid generation may be represented by only one cell, the zygote. Under those circumstances, meiosis takes place immediately after fertilization.

In the alternation of the haploid and diploid generations, the main events are fertilization, which doubles the chromosome number, and meiosis, which halves it. The reduction of the chromosome number during meiosis occurs because the nucleus divides twice, while the chromosomes replicate only once. Another essential characteristic of meiosis is the pairing of homologous chromosomes, which makes their orderly segregation possible.

A further phenomenon, typical of meiosis in most organisms, is the crossing-over or exchange of homologous segments between two of the four chromatids of the paired chromosomes. Crossing-over as such is not an absolute prerequisite for orderly meiotic segregation. For instance, in the *Drosophila* male, normal meiosis occurs, but crossing-over is absent. Another indication of the mutual independence of crossing-over and segregation is provided by *mitotic* crossing-over, which does not result in segregation of homologous chromosomes.

If crossing-over is absent in one sex, it is usually present in the other. Thus, in the *Drosophila* female, crossing-over takes place regularly; its failure leads to the dissociation of paired chromosomes and resultant abnormalities in segregation.

The almost ubiquitous presence of crossing-over in at least one sex per species carries obvious evolutionary benefits. In the offspring, it greatly increases genetic recombination beyond the variability already derived from the independent segregation of maternal and paternal chromosomes in the first meiotic anaphase.

The main features of meiosis—one DNA synthesis, two cell divisions,

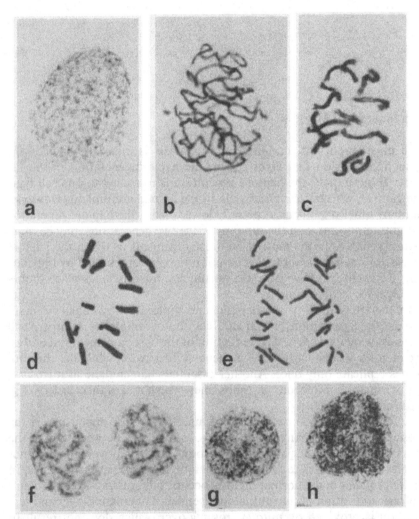

Figure XIV.1. Premeiotic stages and leptotene in pollen mother cells of *Eremurus*. (a) Interphase; (b–c) prophase; (d) metaphase; (e) anaphase; (f) telophase; (g) interphase; (h) leptotene (Feulgen squash).

chromosome pairing, crossing-over, and segregation—are strikingly similar throughout the plant and animal kingdoms. As Darlington puts it, "The lily can tell us what happens in the mouse. The fly can tell us what happens in man." (Darlington and La Cour, 1976, p. 20).

Meiotic Stages

We will follow the premeiotic and meiotic stages in the pollen mother cells of a liliaceous plant, *Eremurus*, which has 14 relatively large chromosomes (Oksala and Therman, 1958; Therman and Sarto, 1977). Before meiosis, the pollen mother cells, which are located in the anther lobes, undergo a number of mitoses. The last of these is shown in Fig. XIV.1.

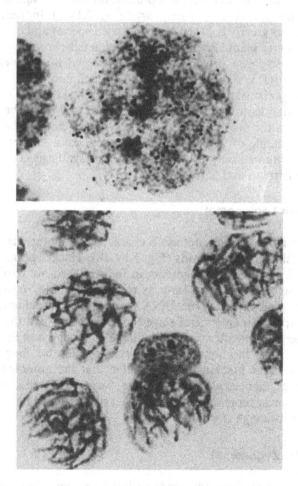

Figure XIV.2. Top: leptotene stage from a human embryonic oocyte undergoing DNA synthesis. Bottom: pachytene bouquet stages in newt (*Triturus vulgaris*) spermatocytes.

Despite a number of claims that the homologous chromosomes show a tendency toward mitotic pairing during the premeiotic stages, no sign of the homologous chromosomes lying side by side is observed. Little, if any, solid evidence exists for chromosome pairing before meiosis (cf. Walters, 1970; John, 1976).

Premeiotic Interphase

Premeiotic contraction has been described in a number of organisms, especially plants (cf. Walters, 1970). Toward the end of the last premeiotic interphase, the chromosomes contract and come to resemble mitotic prophase chromosomes. Later in interphase these contracted chromosomes unravel to form the leptotene threads. No sign of such premeiotic contraction is seen in *Eremurus* (Fig. XIV.1). Indeed, this phenomenon varies greatly between closely related species and even between cells in the same plant. Possibly these contracted chromosomes correspond to the so-called prochromosomes observed in the meiocytes of many animals (cf. Wilson, 1925), including human oocytes.

In some organisms, premeiotic DNA synthesis takes place in the interphase preceding leptotene (cf. Stern and Hotta, 1974). In the liliaceous plant *Trillium*, the period between DNA synthesis and the beginning of meiosis is a couple of weeks; in *Lilium*, a few days. However, in *Eremurus* and in human oocytes (Fig. XIV.2), this synthesis occurs in early leptotene (Therman and Sarto, 1977).

Leptotene (or Leptonema)

The beginning of meiotic prophase is characterized by the chromosomes becoming visible as thin threads (Fig. XIV.1h). Thicker points, or chromomeres, which are coiled segments in the stretched-out chromosomal strand, appear and become clearer as the chromosomes gradually contract. At this stage in *Eremurus*, DNA synthesis can be demonstrated by autoradiography. An ephemeral stage that has been called the "distance pairing" stage follows after typical leptotene (Therman and Sarto, 1977). During this phase the homologues lie side by side for long stretches, apparently without touching. It is possible that the homologous chromosomes sort themselves out in some way during this stage (Fig. XIV.3a); or it may represent early zygotene; synapsis may have started at some points, although it is not visible.

Zygotene (or Zygonema)

In zygotene, the chromosome ends collect at a spot inside the nuclear membrane close to the centrioles, which are outside the nucleus. This movement causes the so-called *bouquet* formation (Figs. XIV.2,4).

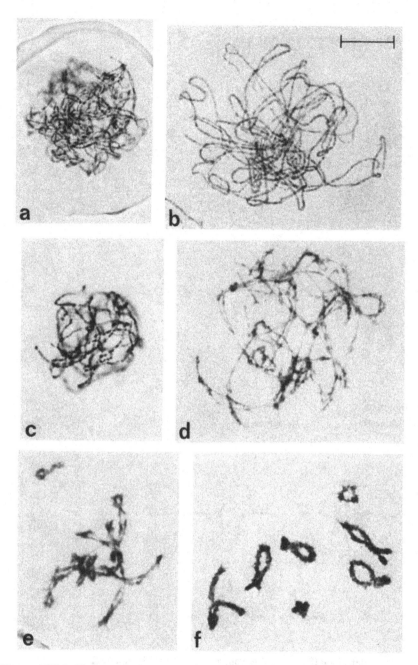

Figure XIV.3. Early meiotic stage in pollen mother cells of *Eremurus*. (a) "Distance pairing" stage; (b) zygotene; (c) pachytene; (d) early diplotene; (e) second contraction in diplotene; (f) late diplotene (Feulgen squash; bar = 15μ) (Therman and Sarto, 1977).

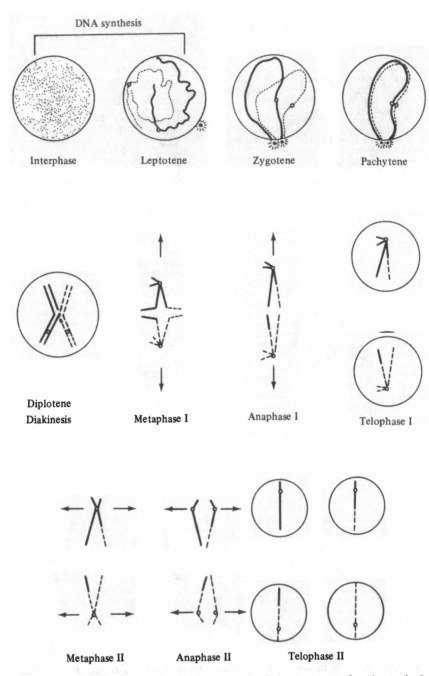

Figure XIV.4. Diagram of meiosis with one pair of chromosomes forming a single chiasma. Note: the chromosomes are *double* after the DNA synthesis, although they have been drawn single in zygotene and pachytene.

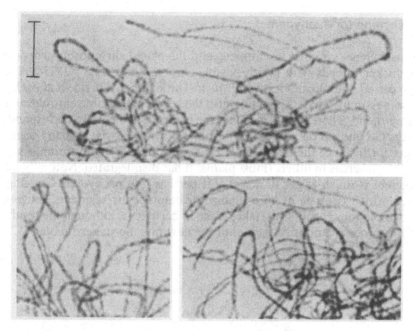

Figure XIV.5. Zygotene in three pollen mother cells of *Eremurus*; paired and unpaired segments are clearly visible (Feulgen squash; bar = 10μ; Therman and Sarto, 1977).

Although the bouquet configuration varies, it seems to be ubiquitous in both plants and animals (Oksala and Therman, 1958). It may be one of the mechanisms that allow the homologous chromosomes to find each other.

The homologous chromosomes pair during zygotene. Parts of the chromosomes are thin leptotene threads, whereas the rest have synapsed to form thicker pachytene-like chromosomes. The points where the pairing has started can be clearly seen (Fig. XIV.5). Now the chromomeres are visible as different-sized "beads" in the chromosome thread.

That paired and unpaired regions are visible in the same chromosome pair indicates that the pairing of homologues starts at several points. Synapsis takes place not only between two homologous chromosomes but between strictly homologous segments. This is convincingly demonstrated by the pairing configuration of two homologues, if one of them has an inversion. To achieve point-by-point pairing in the inverted segment, one of the chromosomes has to form a loop, and such loops are seen regularly. Similarly, if a translocation has taken place between two chromosomes, the corresponding segments of translocated and nontranslocated chromosomes synapse, leading to a characteristic X-shaped figure.

Pachytene (or Pachynema)

In pachytene, synapsis is complete, and the paired chromosomes appear as thicker threads with clearly visible chromomeres (Fig. XIV.3c). The two paired homologues form a *bivalent*. The length of a bivalent at pachytene is estimated to be about one-fifth the length of the same chromosome in leptotene. There is evidence that the chromomere pattern corresponds to the G-bands in the same chromosomes during mitosis. In many organisms, for instance the newt (Fig. XIV.2), the bouquet organization is still visible, whereas in others (most plants) it has disintegrated. Naturally the bouquet is much easier to detect if it persists into pachytene. Crossing-over, which consists of an exchange of homologous segments between two of the four chromatids, takes place in pachytene, although the results of this process cannot be seen until the next meiotic stage, diplotene.

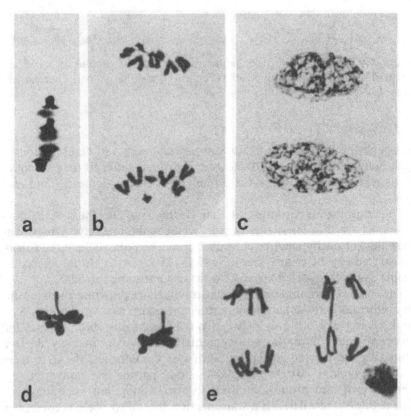

Figure XIV.6. Late meiotic stages in pollen mother cells of *Eremurus*. (a) Side view of metaphase I; (b) anaphase I; (c) interkinesis; (d) side view of metaphase II; (e) II anaphase (Feulgen squash).

Diplotene (or Diplonema)

The shortening and thickening of chromosomes, which take place between leptotene and pachytene, continue in diplotene. The two homologous chromosomes forming a bivalent begin to repel each other until they are held together only at the points of crossing-over or, as these are called in cytological terms, *chiasmata* (singular: *chiasma*) (Fig. XIV.3d–f).

A special type of diplotene chromosome, the lampbrush chromosome, has been studied extensively, especially in the giant oocytes of amphibians, as described in Chapter VI. It is probable that a stage corresponding to the lampbrush diplotene occurs in the meiocytes of many—possibly all— organisms. Indications of this phenomenon are seen in the "hairy" appearance of diplotene chromosomes in both plants and animals. This is especially clear in many grasshoppers.

At some point in mid-diplotene, most of the bivalents collect into one group in the middle of the nucleus (Fig. XIV.3e). The significance of this stage (the second contraction), is not known.

Diakinesis

Chromosome condensation reaches its final stage in diakinesis. The hairy appearance characteristic of diplotene disappears, and the bivalents are smooth and compact. They seem to be attached to the nuclear membrane, whereas until this stage they appeared to lie free inside the nucleus.

Metaphase I

The nuclear membrane vanishes, and during a short prometaphase stage the bivalents collect into a metaphase plate at the midpoint of the spindle, which has now been formed between the two centrioles (Fig. XIV.6a and II.2b). The two centromeres of each bivalent orient themselves toward different poles. The maternal and paternal centromeres are oriented at random, forming the material basis for the independent mendelian segregation of genes.

The shortening and thickening of chromosomes between leptotene and metaphase I are caused largely by the spiralization of the chromosome strand. Compared with mitotic chromosomes, there is an additional supercoil that is beautifully visible in large plant chromosomes, such as those found in *Scilla* or *Tradescantia*.

Anaphase I

In the first meiotic anaphase, the undivided centromeres, with two chromatids attached to each, move to opposite poles.

Telophase I

In the first meiotic telophase, each of the two chromosome groups contains the haploid number of centromeres. If crossing-over has occurred, the two chromatids of a chromosome are not identical. Therefore, these chromosomes are different from their mitotic counterparts.

Interkinesis

In many organisms including *Eremurus*, the first meiotic telophase is followed by a short interphase stage, the *interkinesis* (Fig. XIV.6c). However, during this stage the chromosomes do not replicate and thus have the same basic structure in prophase as they had in the preceding telophase. In those organisms that have no interkinesis (grasshoppers, for example), the chromosome groups of anaphase I become the metaphase plates for the second meiotic division.

Meiotic Division II

In organisms both with and without interkinesis, the second meiotic division is indistinguishable from an ordinary, although haploid, mitosis (Fig. XIV.6d and e). However, the chromosomes in metaphase II often differ from the mitotic chromosomes in the same organism by being shorter and having widely separated chromatids held together only at the centromere. In anaphase II the centromeres divide and move to the poles, with one chromatid attached to each daughter centromere. The end result is four nuclei, each with the haploid chromosome complement of which, however, no two are identical in gene content.

Some Meiotic Features

In some organisms, "terminalization" of chiasmata occurs between diplotene and metaphase I. Chiasmata move toward the chromosome ends and apparently decrease in number by slipping off the ends. This is a different phenomenon from the sliding of chiasmata toward the chromosome ends, which takes place regularly in anaphase I. Because true terminalization of chiasmata, which occurs between diplotene and metaphase I, is probably a much rarer phenomenon than has been assumed (cf. Hultén et al, 1978), many of the older claims should be reinvestigated. As far as we know, terminalization does not have any biological meaning.

After the premeiotic DNA synthesis, the amount of DNA in the nucleus is the same as in the mitotic prophase, 4C. In anaphase I it is reduced to 2C. Since a DNA synthesis never takes place in interkinesis, the DNA amount in each of the four nuclei resulting from meiosis is 1C, which is characteristic of the gametes.

Extensive biochemical studies of meiosis have been made, especially by the group of H. Stern (cf. Stern and Hotta, 1974, 1980). One of the many features distinguishing meiosis from mitosis is that 99.7 percent of the DNA is synthesized in premeiotic interphase or early leptotene. The remaining 0.3 percent of the DNA synthesis occurs during zygotene and seems to play an important role in chromosome pairing. A small amount of repair DNA synthesis occurs in pachytene in connection with crossing-over (cf. Hotta et al, 1977).

The duration of the meiotic stages differs markedly from the mitotic ones in the same organism. Thus, the premeiotic DNA synthesis lasts about twice as long as the mitotic S period. Pachytene lasts 8 days in the male mouse and 16 days in man, whereas the duration of the mitotic prophase can be measured in hours. Diplotene in the oocytes is a special case; in amphibians the lampbrush stage lasts about six months; in human oocytes it may last as long as 45 years (to be described later).

References

Darlington CD, La Cour LF (1976) The handling of chromosomes, 6th edn. Wiley, New York

Hotta Y, Chandley AC, Stern H (1977) Meiotic crossing-over in lily and mouse. Nature 269: 240–242

Hultén M, Luciani JM, Kirton V, et al (1978) The use and limitations of chiasma scoring with reference to human genetic mapping. Cytogenet Cell Genet 22: 37–58

John B (1976) Myths and mechanisms of meiosis. Chromosoma 54: 295–325

Oksala T, Therman E (1958) The polarized stages in the meiosis of liliaceous plants. Chromosoma 9: 505–513

Stern H, Hotta Y (1974) Biochemical controls of meiosis. Annu Rev Genet 7: 37–66

Stern H, Hotta Y (1980) The organization of DNA metabolism during the recombinational phase of meiosis with special reference to humans. Mol Cell Biochem 29: 145–158

Therman E, Sarto GE (1977) Premeiotic and early meiotic stages in the pollen mother cells of *Eremurus* and in human embryonic oocytes. Hum Genet 35: 137–151

Walters MS (1970) Evidence on the time of chromosome pairing from preleptotene spiral stage in *Lillium longiflorum* "Croft". Chromosoma 29: 375–418

Wilson EB (1925) The cell in development and heredity, 3rd edn. MacMillan, New York

XV
Details of Meiosis

Structure of Chiasmata

A diplotene bivalent with one chiasma, as seen in the microscope, can be interpreted according to either the partial *chiasma-type theory* or the *two-plane theory* (Fig. XV.1) (cf. Darlington, 1937). The latter assumes that the paired chromatids on the distal side of a chiasma belong to different chromosomes, one paternal and one maternal. According to the chiasma-type theory, two maternal chromatids and two paternal chromatids remain with each other on both sides of the chiasma: in other words, a chiasma is the result of crossing-over. Observations of normal bivalents do not allow one to distinguish between these two possibilities.

Heteromorphic bivalents provide one of the means of distinguishing between the two theories. If one of the homologues has an added segment resulting from translocation, such a heteromorphic bivalent would have the configuration shown in Fig. XV.1 (top), if the two-plane theory is correct. However, such a configuration has never been observed. Instead, heteromorphic bivalents always have the configuration shown in Fig. XV.1 (bottom), which would be expected on the basis of the chiasma-type theory.

Figure XV.2 illustrates the structure of three bivalents according to the chiasma-type theory. In Fig. XV.2a a diplotene bivalent with one chiasma is shown; in Fig. XV.2b a bivalent with three chiasmata and one chromatid overlap; and in Fig. XV.2c a metaphase bivalent with one chiasma (note that the black chromatid always pairs with another black one, and the interrupted chromatid with a similar one).

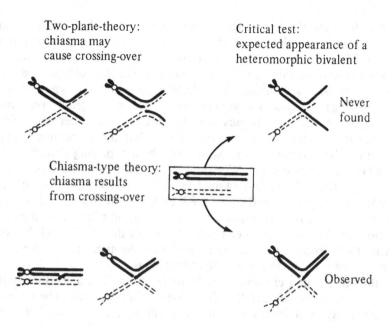

Figure XV.1. Two-plane versus chiasma-type theory as an explanation for the relationship of chiasma and crossing-over.

Number of Chiasmata

The typical bivalent has at least one chiasma. If an organism has chromosomes of different sizes, the larger ones usually show a higher number of chiasmata, sometimes as many as six or eight. At metaphase I, either a bivalent with one chiasma is cross-shaped or a more-or-less terminal chiasma connects the homologues end to end. Two chiasmata often result in a ring-shaped bivalent. In a bivalent with several chiasmata, the loops between them are at right angles to each other so that such a configuration resembles a chain.

Since pairing starts at several points and often several chiasmata are

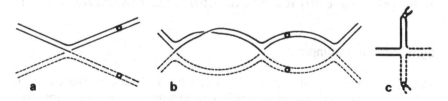

Figure XV.2. Diagram of three bivalents. (a) Diplotene bivalent with one chiasma; (b) bivalent with three chiasmata and one chromatid overlap; (c) metaphase bivalent with one chiasma.

formed between the homologues, one might expect that bivalents and, even more, multivalents would interlock. Actually, interlocking is an exceedingly rare phenomenon. A number of hypotheses have been put forward to explain this. It has been assumed that the "distance pairing" stage or the bouquet formation, or both, represent some type of homologue sorting process before synapsis takes place: this might prevent interlocking when the homologues pair (cf. Oksala and Therman, 1958). Electron-microscopic studies also indicate that the homologous chromosomes align themselves in some way before pairing actually takes place (Westergaard and von Wettstein, 1972). However, the most important clue to the lack of interlocking has been provided by recent electron-microscopic observations (Bojko, 1983; Holm and Rasmussen, 1983; von Wettstein et al, 1984). In zygotene of both male and female meiosis, the interlocked chromosomes break, and afterwards the breaks heal. In other words, the chromosomes are able to "walk through" each other, and therefore no interlocking is visible at metaphase.

Although the first chiasma is presumably formed at a random point, the next one cannot arise in the immediate vicinity but has to be a certain distance away from the first chiasma. This phenomenon is called *chiasma interference*. That crossing-over, which is determined by genetic means, shows a similar interference supplies one of the proofs that chiasmata are the cytological consequence of crossing-over. The correspondence of chiasmata and cross-overs has been demonstrated with autoradiography in the grasshopper (Jones, 1971) and with BrdU in mice (Polani and Crolla, 1982). Furthermore, in corn the total number of cross-overs (as determined by genetic means) was one-half the number of chiasmata, just as expected (Whitehouse, 1969).

Chiasmata may also be more-or-less strictly localized in a certain chromosome region. Thus in some species of the liliaceous genera *Paris*, *Allium*, and *Fritillaria*, as well as in certain grasshoppers, each bivalent has only one chiasma, which is localized near the centromere. In other plants, as well as in some newts, the chiasmata are situated at the ends of chromosomes (cf. Darlington, 1937; John and Lewis, 1965; White, 1973).

In exceptional cases, crossing-over can "go wrong", leading to a dicentric and an acentric chromatid. Such configurations are called U-type exchanges, because they resemble the letter U (cf. Rees and Jones, 1977).

Synaptonemal Complex

The *synaptonemal complexes*, which can be studied with the electron microscope, are essential elements in chromosome pairing and crossing-over. The basic units of the synaptonemal complex are the lateral elements. These structures consist of protein and RNA and arise in the nucleoplasm during the leptotene stage. When chromosome pairing starts, one or both ends of each homologue become attached to the nuclear membrane, and the two homologues become roughly aligned; a lateral element attaches to each of them. The lateral elements pair to form the synaptonemal complex, which includes a central element that completes the synapsis (Fig. XV.3). The formation of a synaptonemal complex between a pair of homologues is initiated at several points. During diplotene the synaptonemal complexes separate from the bivalents and either collect into so-called polycomplexes or disintegrate (cf. Westergaard and von Wettstein, 1972; Luykx, 1974; von Wettstein et al, 1984).

Synaptonemal complexes have been found in all organisms in which chromosome pairing and four-strand crossing-over occur (Westergaard and von Wettstein, 1972). They obviously play an important role in the effective chromosome pairing that is a prerequisite for crossing-over. For example, they are absent in the meiocytes of the *Drosophila* male (an organism in which crossing-over does not take place), but they are present in the female of the same species (in which chiasmata are formed).

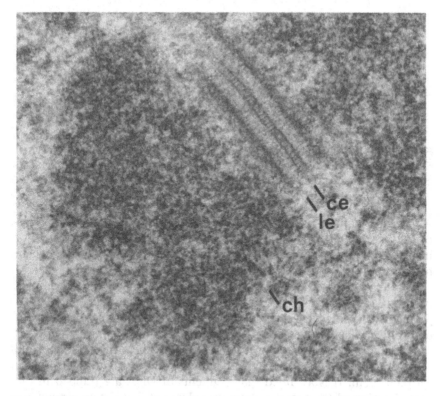

Figure XV.3. Synaptonemal complex from the spermatocyte of the spittle bug *Philaemus* (le = lateral element, ce = central element, ch = chromatin) (courtesy of H Ris).

Figure XV.4. Silver-stained synaptonemal complexes in a spermatocyte of Syrian hamster; the sex chromosomes have a synaptonemal complex between them and more heavily stained differential segments (Pathak, 1983).

Although synaptonemal complexes are apparently needed for crossing-over, their presence does not guarantee the occurrence of this process. For instance, they have been found in the meiocytes of haploid plants where no chiasmata are formed (cf. Gillies, 1975).

Although the discovery of the ubiquity of the synaptonemal complex is important for the interpretation of meiotic phenomena, numerous problems presented by chromosome pairing and crossing-over remain unresolved. The chromosomes are hundreds of times longer than their synaptonemal complexes, and it is unclear how the exact pairing of the corresponding DNA sequences and subsequent crossing-over take place between them. It is assumed that segments of DNA are trapped in the central region of the synaptonemal complex and that the chiasmata are formed between these.

A useful technique to make synaptonemal complexes visible in the light microscope was described by Pathak and Hsu (1979). Figure XV.4 illustrates the synaptonemal complexes in a Syrian hamster spermatocyte revealed with this silver-staining technique. As in many mammals, the X and the Y chromosomes have a pairing segment in common, and in addition each has a differential segment.

New light has been thrown on the mechanics of crossing-over by the electron microscopic discovery of *recombination nodules*. These are electron-dense structures that are associated with the central element of the synaptonemal complex. They arise in zygotene and transform during pachytene into bridge-like structures, called *bars*. Crossing-over takes place during this transformation, which has been inferred from the agreement between crossing-over and the number, distribution, and structure of the nodules and bars. Recombination nodules are thought to be a prerequisite for crossing-over, as first suggested by Carpenter (1975) (see also Holm and Rasmussen, 1981, 1983; von Wettstein et al, 1984).

Meiotic Behavior of More Than Two Homologous Chromosomes

Meiosis in polyploids has been analyzed extensively, especially in plants. It is of little significance in human cytogenetics, since the only polyploids known—triploids and tetraploids—as a rule die prenatally and survive birth only exceptionally. However, the early meiotic stages in oocytes of a triploid human fetus have been analyzed by Luciani et al (1978).

Women with Down's syndrome (21 trisomy) are the only individuals with autosomal trisomy known to have reproduced; their offspring consist of children with Down's syndrome and normal children in a ratio that does not differ significantly from the expected 1:1 (cf. Hamerton, 1971, p. 214).

Three homologous chromosomes usually pair to form a *trivalent* in meiosis (Fig. XV.5). However, at any one point only two of them synapse (cf. Darlington, 1937). As shown in Fig. XV.5, two chromosomes (the thick and the interrupted line) start to pair, then there is an exchange of partners (the thick pairs with the thin), and at the second exchange the first two chromosomes (the thick and the interrupted line) resume pairing. If chiasmata are formed in the three paired segments, the diplotene configuration shown in Fig. XV.5 results. An alternate centromere orientation gives rise to the metaphase I trivalent in Fig. XV.5. Depending on the number and length of the paired segments and the number and location of chiasmata, trivalents come in a variety of shapes. Three homologues may also form a bivalent and a univalent. Of the three homologues, usually two go to one pole and one to the other in anaphase I. This type of segregation is called *secondary nondisjunction*.

Four homologous chromosomes may form a *quadrivalent*, or two bivalents, or a trivalent and a univalent. Even more possibilities exist in

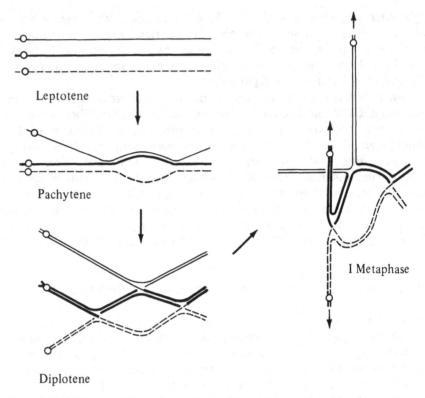

Figure XV.5. Pairing of three homologous chromosomes. Note: the chromosomes are *double* in pachytene although they are drawn single.

higher polyploids, such as hexaploids, in which chains or rings of six chromosomes are observed. Because meiotic segregation in polyploids is often irregular, the gametes may receive inviable chromosome constitutions, resulting in partial sterility.

Human Meiosis

In the human male, as in the males of most plants and animals, the four meiotic products (spermatids) form four functional gametes. Meiosis in the oocyte results in one functional egg cell and three very small cells. In animals these are called polar bodies and degenerate, whereas in plants the three nuclei are included in the structure of the embryo sac.

From a cytologist's point of view, meiosis in human oocytes complements that in spermatocytes in that the meiotic prophase stages have been studied—and are much clearer—in the female, whereas our information on stages from diplotene onward comes mainly from the male.

Premeiotic and Early Meiotic Stages in Man

The early meiotic stages in the human female are found in 12- to 16-week-old embryos. The last premeiotic mitosis, in which the metaphase chromosomes appear short and thick, is followed by an interphase in which most or all of each chromosome remains condensed as a chromocenter. These chromocenters unravel to form the leptotene threads. The meiotic DNA synthesis occurs in early leptotene (Fig. XIV.2), as in *Eremurus* pollen mother cells. Between leptotene and zygotene there is a short "distance pairing" stage (Therman and Sarto, 1977)—again as in *Eremurus*—during which the homologues appear to lie side by side but at a distance (Fig. XV.6).

The diplotene stage in human oocytes is a very long phase, lasting from the embryonic stage until the egg cell is released in the adult female. This type of diplotene, called the *dictyotene*, represents a period of enormous growth of the cell. The chromosomes are greatly extended and invisible with most cytological techniques. Dictyotene corresponds to the lampbrush stage, and in addition to the normal nucleoli 15–20 extra nucleoli are formed, which seem to be attached to the heterochromatic centric regions of the chromosomes, especially of chromosome 9 (Stahl et al, 1975). A similar diffuse diplotene stage is also described in the spermatocytes of insects and many other organisms.

The bivalent, consisting of the two X chromosomes, behaves in the

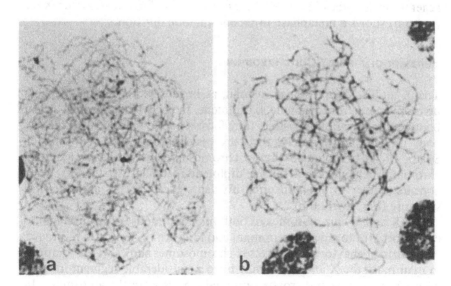

Figure XV.6. Meiotic stages in human oocytes. (a) "Distance pairing" stage; (b) zygotene (Therman and Sarto, 1977). (Note that the X chromosomes are not distinguishable from the autosomes).

oocytes in the same way as the autosomal bivalents do (Fig. XV.6), that is, without showing any heteropycnotic condensation. In this respect the X chromosomes behave differently in the oocytes from female somatic cells, where one of them acts heterochromatically. In the spermatocytes of the male, both the X and the Y chromosomes are heteropycnotically condensed in the meiotic prophase from zygotene to diplotene.

Meiosis in Human Spermatocytes

Figure XV.7 (top) illustrates a spermatogonial metaphase in which the spiral structure of the chromosomes is clearly visible. In Fig. XV.7 (bottom) a pachytene stage with an XY body is shown; the bouquet structure has been destroyed by squashing. The mean chiasma frequency in human spermatocytes, as determined from diakinesis and metaphase I stages, is 51 (SD 3.9) (Hultén, 1974). The number of chiasmata in individual cells varies from 43 to 60. As seen in Fig. XV.8 (top), the largest human bivalents usually have four chiasmata; the medium-sized, two to three; and the smallest ones, one each. In many organisms the chiasma frequency in the oocytes and the spermatocytes is significantly different. Whether this applies to man also is not known, since so little information exists about the later meiotic stages in the female. However, there is evidence of unequal recombinant frequencies between the sexes.

   The second meiotic division in the spermatocytes follows the usual scheme, as described in the pollen mother cells of *Eremurus*. Figure XV.8 (bottom) shows a metaphase in second meiotic division.

Behavior of X and Y Chromosomes

The X and Y chromosomes in man, as in most other animals, appear heteropycnotic from zygotene to diplotene. In pachytene, in which a clear bouquet organization is visible, they form a paired clump called the XY body or (less appropriately) the "sex vesicle" (Fig. XV.7, bottom) (cf. Solari, 1974). In diplotene, the sex chromosomes cease to show a heterochromatic behavior, and during diplotene-metaphase I, the XY body appears similar to the autosomal bivalents except for its asymmetrical structure (Fig. XV.8, top).

   The X and Y chromosomes of both man and mouse are paired end to end, short arm to short arm in man, and long arm to long arm in mouse. The pairing behavior of the two sex chromosomes shows wide variations in mammals: the X and the Y may have a considerable segment in common where pairing and crossing-over take place as in the autosomes. In addition, they have a differential segment in which the sex determinants are situated. A real chiasma between the sex chromosomes has been observed, for instance, in several hamster species representing different

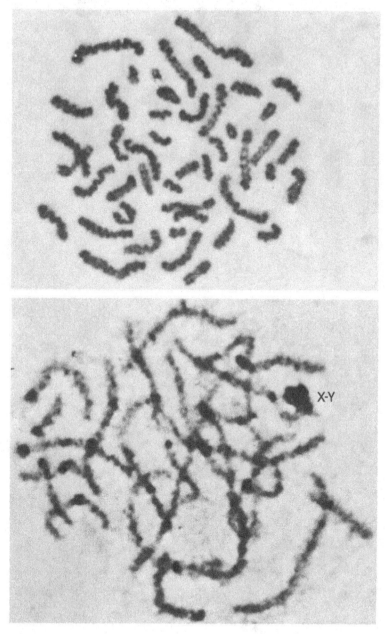

Figure XV.7. Top: Metaphase from a human spermatogonial cell. Bottom: Pachytene from a human spermatocyte with an XY body (the bouquet arrangement has been destroyed by squashing) (courtesy of M Hultén).

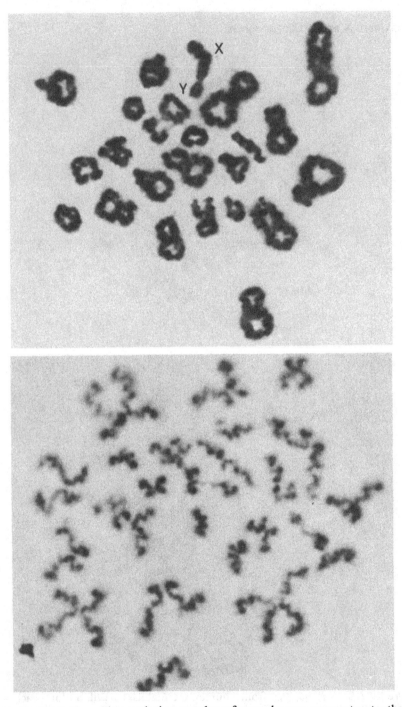

Figure XV.8. Top: First meiotic metaphase from a human spermatocyte; the X and the Y are attached, short arm to short arm. Bottom: Second meiotic metaphase from a human spermatocyte (courtesy of M Hultén).

genera (cf. Solari, 1974). In both mouse and man the pairing segment is small, and a short synaptonemal complex is formed (Solari, 1980).

However, no chiasma has been observed between the human sex chromosomes. How this is to be reconciled with the claim of Holm and Rasmussen (1983) that crossing-over occurs between them in 75% of the cases, which they have concluded from the presence of recombination nodules, is not clear.

References

Bojko M (1983) Human meiosis VIII. Chromosome pairing and formation of the synaptonemal complex in oocytes. Carlsberg Res Commun 48: 457–483

Carpenter ATC (1975) Electron microscopy of meiosis in Drosophila melanogaster. II. The recombination nodule—a recombination-associated structure at pachytene? Proc Natl Acad Sci USA 72: 3186–3189

Darlington CD (1937) Recent advances in cytology, 2nd edn. Churchill, London

Gillies CB (1975) Synaptonemal complex and chromosome structure. Annu Rev Genet 9: 91–109

Hamerton JL (1971) Human cytogenetics, Vol II. Academic, New York

Holm PB, Rasmussen SW (1981) Chromosome pairing, crossing over, chiasma formation and disjunction as revealed by three dimensional reconstructions. In: Schweiger HG (ed) International cell biology 1980–1981. Springer, Berlin, pp 195–204

Holm PB, Rasmussen SW (1983) Human meiosis VI. Crossing over in human spermatocytes. Carlsberg Res Commun 48: 385–413

Hultén M (1974) Chiasma distribution at diakinesis in the normal human male. Hereditas 76: 55–78

John B, Lewis KR (1965) The meiotic system. In: Protoplasmatologia, Vol VI, Fl. Springer, New York

Jones GH (1971) The analysis of exchanges in tritium-labelled meiotic chromosomes. Chromosoma 34: 367–382

Luciani JM, Devictor M, Boué J, et al (1978) The meiotic behavior of triploidy in a human 69,XXX fetus. Cytogenet Cell Genet 20: 226–231

Luykx P (1974) The organization of meiotic chromosomes. In: Busch H (ed) The cell nucleus, Vol II. Academic, New York, pp 163–207

Oksala T, Therman E (1958) The polarized stages in the meiosis of liliaceous plants. Chromosoma 9: 505–513

Pathak S (1983) The behavior of X chromosomes during mitosis and meiosis. In: Sandberg AA (ed) Cytogenetics of the mammalian X chromosome. Part A: Basic mechanisms of X chromosome behavior. Liss, New York, pp 67- 106

Pathak S, Hsu TC (1979) Silver-stained structures in mammalian meiotic prophase. Chromosoma 70: 195–203

Polani PE, Crolla JA (1982) Experiments on female mammalian meiosis. In: Crosignani PG, Rubin BL (eds) Genetic control of gamete production and function. Academic, London, pp 171–186

Rees H, Jones RN (1977) Chromosome genetics. University Park Press, Baltimore

Solari AJ (1974) The behavior of the XY pair in mammals. In: Bourne GH, Danielli JF (eds) International review of cytology, Vol 38. Academic, New York, pp 273–317

Solari AJ (1980) Synaptonemal complexes and associated structures in micro-spread human spermatocytes. Chromosoma 81: 315–337

Stahl A, Luciani JM, Devictor M, et al (1975) Constitutive heterochromatin and micronucleoli in the human oocyte at the diplotene stage. Humangenetik 26: 315–327

Therman E, Sarto GE (1977) Premeiotic and early meiotic stages in the pollen mother cells of *Eremurus* and in human embryonic oocytes. Hum Genet 35: 137–151

Westergaard M, Wettstein D von (1972) The synaptinemal complex. Annu Rev Genet 6: 71–110

Wettstein D von, Rasmussen SW, Holm PB (1984) The synaptonemal complex in genetic segregation. Annu Rev Genet 18: 331–431

White MJD (1973) Animal cytology and evolution, 3rd edn. University Press, Cambridge, England

Whitehouse HLK (1969) Towards an understanding of the mechanism of heredity. Arnold, London

XVI
Meiotic Abnormalities

Nondisjunction of Autosomes

Meiosis is a much more complicated chain of events than mitosis. Consequently the chances of failure during some part of the process are also more numerous. Most meiotic irregularities lead to nondisjunction or to chromosome loss, which in turn results in the formation of gametes with aneuploid chromosome numbers. Nondisjunction refers to any process resulting in two chromosomes that ought to segregate to opposite poles going to the same pole. Meiosis may also be disrupted at various points resulting in degeneration of the meiocytes.

Some of the possible meiotic aberrations are illustrated in Fig. XVI.1, which also shows the fate of two allelic genes, A and a. In the first case, the two chromosomes of a bivalent have failed to disjoin and therefore go to the same pole. In the second meiotic division, the two homologues divide normally, resulting in the formation of two disomic and two nullosomic gametes. In the following example, the chromosomes have paired normally but no chiasma has formed. Consequently the paired homologues fall apart and appear as univalents.

Univalents may behave in different ways; they may drift at random to the two poles in the first division and divide regularly in the second. Alternatively, they may divide mitotically in anaphase I and, being single chromatids in the second division, they cannot divide any more and so drift at random to the poles or misdivide. Similar behavior is also exhibited by univalents if they result from an original failure to pair in zygotene. One possibility is that the drifting univalents do not reach the poles but remain as laggards and are lost. Univalents also often undergo misdivision.

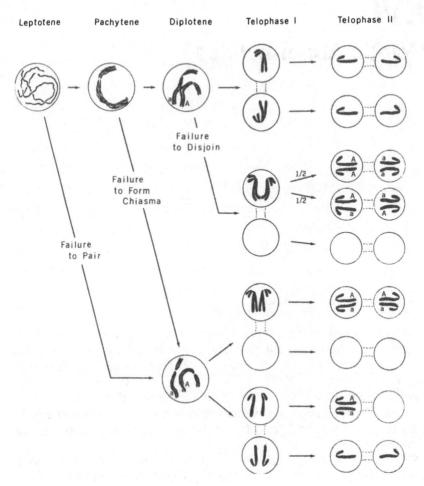

Figure XVI.1. Diagram of the processes resulting in meiotic nondisjunction (see text) (Patau, 1963).

Nondisjunction of Sex Chromosomes

Since the XX bivalent in mammalian oocytes behaves like the auto-somes, the two X chromosomes in the female exhibit the same irregular-ities in pairing and segregation as the rest of the chromosomes.

The behavior of the X and the Y chromosomes in the male, on the other hand, differs from the autosomes. Although a short synaptonemal complex is formed between the two sex chromosomes, their end-to-end attachment seems to be different in kind from a typical terminal chiasma. This is confirmed by the fact that the X and the Y are found much more often as univalents than are the small autosomes with one terminal chiasma.

In a study of meiosis in 53 normal men, the mean fraction of spermatocytes in which the X and the Y were unpaired was 3.2 percent (McDermott, 1971). Much higher frequencies of separation of the sex chromosomes have also been reported (cf. Solari, 1974, p. 304), but many of these earlier claims sound unrealistic. In normal mice, 8 to 10 percent of the spermatocytes are observed to have a separate X and Y (cf. Rapp et al, 1977).

Misdivision of the Centromere

Misdivision or *fission* of the centromere may happen during mitosis or meiosis. However, since it probably occurs much more often during meiosis, it is discussed at this point. In misdivision or fission, the centromere divides crosswise, separating the two chromosome arms instead of the two chromatids as in normal division (Fig. XVI.2).

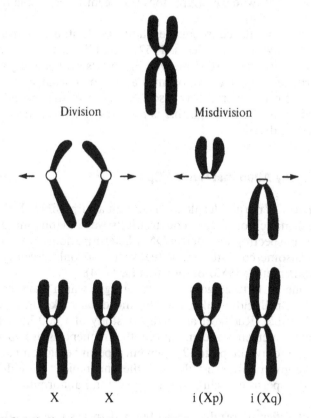

Division   Misdivision

X   X   i (Xp)   i (Xq)

Figure XVI.2. Misdivision of the human X chromosome resulting in the formation of a long-arm isochromosome i(Xq) and a short-arm isochromosome i(Xp) (presumably inviable).

If misdivision takes place during the stages from S to anaphase when the chromosomes are double, the result is two isochromosomes. If a single chromatid misdivides during the stages from anaphase to S, two telocentric chromosomes arise. In three humans, chromosomes 4, 7, and 10 have undergone fission, respectively, resulting in two telocentrics in each cell of the person affected (cf. Therman et al, 1981).

An abnormal child with symptoms of 13 trisomy was found to have two cell lines, one with a telocentric chromosome 13 and another with i(13q) (Therman et al, 1963). Other similar cases involving chromosomes 13, 21, and the X have since been described (cf. Therman et al, 1981). In these cases either the telocentric or the isochromosome has arisen through misdivision of a normal chromosome and the other through further misdivision of the first abnormal chromosome (once a centromere misdivides, it often continues to be unstable).

However, misdivision most often affects univalents during the first meiotic division. In humans the most frequent result of misdivision is a monocentric i(Xq), whereas i(Xp) does not seem to be viable (cf. Therman and Patau, 1974).

Misdivision and its consequences have been studied especially in plants (cf. Darlington, 1939; Sears, 1952). In addition to the centromere dividing cross-wise, it can separate into four parts, which will give rise to two long-arm telocentric chromosomes and two short-arm telocentrics. Misdivision of centromeres may also take place in the second meiotic division or in mitosis, although this is much rarer than its occurrence in the first meiotic division.

Chromosomally Abnormal Human Sperm

That the sperm of normal fertile men contain a surprisingly high proportion of abnormal chromosome constitutions was obvious more than 30 years ago from spectrophotometric DNA determinations. However, what these chromosome constitutions actually are has only recently become clear (cf. Martin et al, 1983; Brandriff et al, 1984).

A technique to fertilize golden hamster eggs which have been freed from the zona pellucida makes the chromosomal make-up in the male pronucleus visible (Rudak et al, 1978). A study of 1,000 human sperm showed that 5.2 percent were aneuploid, the frequencies of hyperhaploid (2.4 percent) and hypohaploid (2.7 percent) sperm being comparable. All chromosome groups were involved in the aneuploidy. In addition, 3.3 percent of the sperm had chromosome structural abnormalities (Martin et al, 1983).

Somewhat different results were obtained in a similar study of 909 sperm (Brandriff et al, 1984), of which 1.6 percent showed aneuploidy and 6.5 percent structural abnormalities.

Male Infertility

Male infertility in mammals, which seems to have many more causes than infertility in females, is usually correlated with a low number of sperm in the semen (oligospermia) or with their total lack (azotospermia).

As a rule these conditions come about through meiotic abnormalities which interrupt the development of spermatocytes at some stage, usually during the first meiotic division. For instance, Koulischer et al (1982) reported meiotic abnormalities in 13.2 percent of infertile men.

A great variety of chromosomal anomalies cause male sterility. Thus, depending on patient selection, from 5 to 15 percent of the males sampled in infertility clinics have abnormal chromosome constitutions. For instance, the incidence of balanced translocations, including Robertsonian translocations, is 8.9 per thousand in infertile men as compared with 1.4 per thousand in the newborn (Zuffardi and Tiepolo, 1982). Translocations between two autosomes, between the X or the Y and an autosome, and between an X and a Y cause sterility both in mouse (Searle, 1982) and human males (cf. Chandley, 1982).

Furthermore practically all abnormal sex chromosome constitutions lead to male sterility (cf. Chandley, 1982). For instance, in males with Klinefelter's syndrome (47,XXY), the atrophied seminiferous tubules contain only Sertoli cells, such males always being sterile. On the other hand males with 47,XYY range from fully fertile to sterile. In many of the fertile representatives of this group, one of the Y chromosomes is lost from the germline. In others the two Y chromosomes pair with each other, leaving the X unpaired. Sterility seems to result when one Y pairs with the X chromosome, the other Y remaining unpaired (cf. Forejt, 1982).

Interestingly, in mouse and human spermatocytes with normal sex chromosomes, the pairing of the X and the Y chromosomes seems to be a prerequisite for the completion of meiosis and thus for the formation of spermatozoa (cf. Rapp et al, 1977).

It has been assumed that the damaging effect of most chromosome anomalies on the course of meiosis involves their interference with the pairing of the X and Y chromosomes and with their heterochromatinization during meiosis (cf. Forejt, 1982).

Mutant genes that affect male meiosis are known in many organisms, both plants and animals. Several lines of evidence indicate that similar genes which prevent chromosome pairing or recombination also exist in humans (cf. Chandley, 1982).

Environmental Causes of Meiotic Nondisjunction

The causes of nondisjunction may be environmental or genetic. Experiments conducted on plants and animals reveal that a variety of environmental agents, both physical and chemical, affect meiosis. For example, heat shocks prevent chromosome pairing and lead to a totally irregular meiosis. The same phenomenon is caused by cold treatment in mouse embryos (Karp and Smith, 1975). Similarly, ionizing radiation increases nondisjunction in *Drosophila* (cf. Uchida, 1977).

Many substances are known to cause chromosome breakage (cf. Kihlman, 1966; Rieger and Michaelis, 1967). Although less attention has been paid to nondisjunction, most of the chromosome-breaking substances probably also give rise to occasional abnormal segregation in mitosis or meiosis. For instance, alkylating leukemia drugs increase meiotic nondisjunction in mice (Röhrborn, 1971; Vogel et al, 1971).

Another class of substances, of which the most widely used is the alkaloid colchicine, specifically destroys the spindle structure. Instead of collecting into a metaphase plate, the chromosomes are scattered around the cell, which in milder cases leads to nondisjunction and in extreme cases to restitution, giving rise to a tetraploid cell. As early as 1939, Levan showed that colchicine interfered with chromosome pairing in the meiosis of the onion. Shepard et al (1974) found that colchicine largely destroys the lateral elements of the synaptonemal complex. As a result, only limited stretches of chromosomes pair effectively and form chiasmata. Interestingly the decrease in chiasma formation affects preferentially the bivalents which normally have one chiasma, these chromosomes thus remaining as univalents, whereas the chiasma frequency is not decreased in bivalents with four chiasmata (Bennett et al, 1979). The cytological effects of colchicine should be taken into consideration when this alkaloid is used as a gout treatment for patients of reproductive age.

We have much less definitive information about the causes of meiotic nondisjunction in man than in plants and animals. However, the results of this process—aneuploid sperm, trisomic and monosomic offspring, and chromosomally abnormal spontaneous abortions—demonstrate its existence. It is not far-fetched to assume that the same agents that cause nondisjunction in other organisms also affect human meiosis.

Maternal Age

A well-established cause of nondisjunction in man is increased maternal age, which, as far as the gametes are concerned, can be classified as an environmental influence. At the maternal age of 20 years, the incidence of 21-trisomic children is 0.4 per 1000 newborns; for women age 45 years and over, the risk increases to 17 per 1000 newborns (cf. Hassold and Jacobs, 1984). A similar maternal age effect is found for 18- and 13-trisomic children as well as for offspring with XXX and XXY sex chromosome constitutions. However, this rule does not apply to Turner's syndrome with the 45,X chromosome constitution. Spontaneous abortions with aneuploid chromosome numbers also occur more frequently in older women (cf. Carr, 1971; Hassold et al, 1980). It is estimated that, if women over 35 years of age refrained from having offspring, the incidence of children with abnormal chromosome numbers would decrease by one-third to one-half.

However, with chromosome banding techniques and more refined statistical methods, it is possible to demonstrate that the earlier belief that the father's age played no role in the birth of aneuploid children is untrue. Indentification of the individual 21 chromosomes with banding techniques shows that nondisjunction for this chromosome also occurs in the father; possibly one-third of 21-trisomic children owe this effect to paternal nondisjunction (Wagenbichler et al, 1976; Uchida, 1977; Mattei et al, 1979), and from age 55 onward their frequency increases with paternal age (Stene et al, 1977). Juberg and Mowrey (1983) have pooled the data from relevant studies through 1982, which show that maternal origin accounts for 80 percent of 21 trisomy cases and paternal origin for 20 percent. Nondisjunction in the mother takes place during the first meiotic division in 80 percent of the cases and during the second meiotic division in 20 percent; the corresponding percentages in the father are 60 and 40.

It has also been claimed that the frequency of chromosomally abnormal embryos increases more with advancing paternal age when the mother is older than if she is young (Stene et al, 1981).

In mouse oocytes the frequency of chiasmata is decreased and the incidence of univalents increased with increasing age of the female (Henderson and Edwards, 1968). These results were confirmed by Luthardt et al (1973), who also observed that the univalent formation was nonrandom, involving mainly the smaller chromosomes. This finding agrees with observations on many organisms that the smallest chromosomes of the complement with at most one chiasma are most likely to remain as univalents.

The mechanism by which the age of the female acts on meiosis is not known. Crossing-over takes place during pachytene in embryonic oocytes, and it seems highly improbable that the number of chiasmata would decrease even if the diplotene stage lasted for a very long time. A more plausible explanation is that there is a gradient in the ovary, and the egg cells that are released earlier have more chiasmata than those released later. Furthermore, hormonal and other physiological changes in an aging woman might also interfere with the normal course of meiosis.

The idea has also been advanced that in older women there is a decreased selection against abnormal embryos, which in younger women end up as spontaneous abortions (Aymé and Lippman-Hand, 1982). However, since abortions increase with age, this hypothesis seems improbable.

Another factor that increases nondisjunction in women is preconception exposure to ionizing radiation (Uchida, 1977). Although in a couple of studies no such effect could be established, the accumulated evidence from several others seems to speak for the assumption that radiation must be taken seriously as a possible cause for the birth of human trisomics. Radiation seems to increase nondisjunction, especially in older women and in older female mice (Uchida, 1977). However, no increase in nondisjunction has been found in the Hiroshima and Nagasaki victims.

Genetic Causes of Nondisjunction

In many organisms, for instance in *Drosophila* and maize, genes are known that upset the normal course of meiosis by affecting either synapsis or chiasma formation. In man, a number of abnormal individuals are described who show aneuploidy for more than one chromosome. This fact naturally reflects a more serious disturbance of meiosis than simple nondisjunction. For example, both an extra 21 and an extra 18 were found in an infant who died 8 h after birth (Gagnon et al, 1961). Other combinations are trisomy 21 and an XXX or XXY sex chromosome constitution, or an extra 21 combined with a missing X chromosome (cf. Mikkelson, 1971). Other types of double aneuploidy are occasionally found (cf. Tuck et al, 1984), and even a child with 49,XXYY,+18 has been described (Webb et al, 1984).

The occurrence of more than one child with an aneuploid chromosome constitution in the same sibship has led to the conclusion that such families may have an inherent tendency to nondisjunction. Thus the second D-trisomic child ever described had a sister with 45,X chromosome constitution (Therman et al, 1961). However, the most convincing evidence that some families have a genetic disposition to nondisjunction is from Hecht et al (1964). The study started with 60 families with either an 18-trisomic or D-trisomic offspring. Three of the 18-trisomic children had siblings with 21 trisomy as compared with an expectation of 0.15 mongoloid children for the same group—a highly significant statistical difference. In addition, one of the children with D-trisomy had an uncle with 21 trisomy.

Genetic factors are probably also responsible for the observation that if a couple has one trisomic child, the risk of having another increases some ten times compared with the general population. This probably depends on some families having a much higher risk, whereas in others the risk is not increased. An example of a high-risk family was studied in our laboratory. The couple had three children with Down's syndrome and the father's brother also had a Down's syndrome child. The two fathers had normal chromosomes and the affected children "normal" 47,+21 chromosome constitutions.

Origin of Diploid Gametes

Tetraploid and triploid abortions, and even a few liveborn children, demonstrate that unreduced gametes are formed and are occasionally able to function. Further, spectrophotometric measurements show that some sperms contain double the usual amount of DNA. Such gametes owe their origin mainly to two processes. The same agents causing nondisjunction may disturb meiosis to the extent that a restitution nucleus, which contains the unreduced diploid complement, is formed in the first meiotic division, or endoreduplication in a gonial cell may lead to one or more tetraploid meiocytes. If such a cell subsequently undergoes normal meiosis, it may give rise to diploid gametes.

By using banding techniques to extract informative data on 10 triploid abortuses, Kajii and Niikawa (1977) found that one originated in maternal first meiotic division; five apparently resulted from two sperms fertilizing the same egg; two owed their origin to an aberration in either the second paternal division or the first mitotic division; and the last two were of undefined paternal origin. One tetraploid abortion obviously came about by the suppression of the first cleavage division. Similar results were obtained by Jacobs et al (1978) in a study on 26 triploid abortions. To quote their paper (p. 56): "The best fit for the data using a maximum-likelihood method was that 66.4% of the triploids were the result of dispermy, 23.6% the result of fertilization of a haploid ovum by a diploid sperm formed by failure of the first meiotic division in the male, and 10% the result of a diploid egg formed by failure of the first maternal meiotic division." These observations have been confirmed by Uchida and Freeman (1985).

References

Aymé S, Lippman-Hand A (1982) Maternal-age effect in aneuploidy: Does altered embryonic selection play a role? Am J Hum Genet 34: 558–565
Bennett MD, Toledo LA, Stern H (1979) The effect of colchicine on meiosis in *Lilium speciosum* cv. "Rosemede". Chromosoma 72: 175–189
Brandriff B, Gordon L, Ashworth L, et al (1984) Chromosomal abnormalities in human sperm: Comparisons among four healthy men. Hum Genet 66: 193–201
Carr DH (1971) Chromosomes and abortion. In: Harris H, Hirschhorn K (eds) Advances in human genetics. Plenum, New York
Chandley AC (1982) Normal and abnormal meiosis in man and other mammals. In: Crosignani PG, Rubin BL (eds) Genetic control of gamete production and function. Academic, London, pp 229–237
Darlington CD (1939) Misdivision and the genetics of the centromere. J Genet 37: 341–364
Forejt J (1982) X-Y involvement in male sterility caused by autosome translo-

cations—a hypothesis. In: Crosignani PG, Rubin BL (eds) Genetic control of gamete production and function. Academic, London, pp 135–151

Gagnon J, Katyk-Longtin N, de Groot JA, et al (1961) Double trisomie autosomique a 48 chromosomes (21 + 18). L'Union Med Canada 90: 1–7

Hassold TJ, Jacobs PA (1984) Trisomy in man. Annu Rev Genet 18: 69–97

Hassold TJ, Jacobs P, Kline J, et al (1980) Effect of maternal age on autosomal trisomies. Ann Hum Genet 44: 29–36

Hecht F, Bryant JS, Gruber D, et al (1964) The nonrandomness of chromosomal abnormalities. New Eng J Med 271: 1081–1086

Henderson SA, Edwards RG (1968) Chiasma frequency and maternal age in mammals. Nature 218: 22–28

Jacobs PA, Angell RR, Buchanan IM, et al (1978) The origin of human triploids. Ann Hum Genet 42: 49–57

Juberg RC, Mowrey PN (1983) Origin of nondisjunction in trisomy 21 syndrome: All studies compiled, parental age analysis, and international comparisons. Am J Med Genet 16: 111–116

Kajii T, Niikawa N (1977) Origin of triploidy and tetraploidy in man: 11 cases with chromosome markers. Cytogenet Cell Genet 18: 109–125

Karp LE, Smith WD (1975) Experimental production of aneuploidy in mouse oocytes. Gynecol Invest 6: 337–341

Kihlman BA (1966) Actions of chemicals on dividing cells. Prentice-Hall, Englewood Cliffs, New Jersey

Koulischer L, Schoysman R, Gillerot Y, et al (1982) Meiotic chromosome studies in human male infertility. In: Crosignani PG, Rubin BL (eds) Genetic control of gamete production and function. Academic, London, pp 239–260

Levan A (1939) The effect of colchicine on meiosis in *Allium*. Hereditas 25: 9–26

Luthardt FW, Palmer CG, Yu P-L (1973) Chiasma and univalent frequencies in aging female mice. Cytogenet Cell Genet 12: 68–79

Martin RH, Balkan W, Burns K, et al (1983) The chromosome constitution of 1000 human spermatozoa. Hum Genet 63: 305–309

Mattei JF, Mattei MG, Ayme S, et al (1979) Origin of the extra chromosome in trisomy 21. Hum Genet 46: 107–110

McDermott A (1971) Human male meiosis. Can J Genet Cytol 13: 536–549

Mikkelsen M (1971) Down's syndrome: Current stage of cytogenetic research. Humangenetik 12: 1–28

Patau K (1963) The origin of chromosomal abnormalities. Pathol Biol 11: 1163–1170

Rapp M, Therman E, Denniston C (1977) Nonpairing of the X and Y chromosomes in the spermatocytes of BDF₁ mice. Cytogenet Cell Genet 19: 85–93

Rieger R, Michaelis A (1967) Die Chromosomenmutationen. Gustav Fischer, Jena

Röhrborn G (1971) Chromosome aberrations in oogenesis and embryogenesis of mammals and man. Arch Toxikol 28: 115–119

Rudak E, Jacobs PA, Yanagimachi R (1978) Direct analysis of the chromosome constitution of human spermatozoa. Nature 274: 911–913

Searle AG (1982) The genetics of sterility in the mouse. In: Crosignani PG, Rubin BL (eds) Genetic control of gamete production and function. Academic, London, pp 93–114

Sears, ER (1952) Misdivision of univalents in common wheat. Chromosoma 4: 535–550

Shepard J, Boothroyd ER, Stern H (1974) The effect of colchicine on synapsis and chiasma formation in microsporocytes of *Lilium*. Chromosoma 44: 423–437

Solari AJ (1974) The behavior of the XY pair in mammals. In: Bourne GH,

Danielli JF (eds) International review of cytology, Vol 38. Academic, New York, pp 273–317

Stene J, Fischer G, Stene E (1977) Paternal age effect in Down's syndrome. Ann Hum Genet 40: 299–306

Stene J, Stene E, Stengel-Rutkowski S, et al (1981) Paternal age and Down's syndrome. Data from prenatal diagnoses (DFG). Hum Genet 59: 119–124

Therman E, Patau K (1974) Abnormal X chromosomes in man: Origin, behavior and effects. Humangenetik 25: 1–16

Therman E, Patau K, DeMars RI, et al (1963) Iso/telo-D_1 mosaicism in a child with an incomplete D_1 trisomy syndrome. Portugal Acta Biol 7: 211–224

Therman E, Patau K, Smith DW, et al (1961) The D trisomy syndrome and XO gonadal dysgenesis in two sisters. Am J Hum Genet 13: 193–204

Therman E, Sarto GE, DeMars RI (1981) The origin of telocentric chromosomes in man: A girl with tel(Xq). Hum Genet 57: 104–107

Tuck CM, Bennett JW, Varela M (1984) Down's syndrome and familial aneuploidy. In: Berg, JM (ed) Perspectives and progress in mental retardation II. University Park Press, Baltimore, pp 167–180

Uchida IA (1977) Maternal radiation and trisomy 21. In: Hook EB, Porter IH (eds) Population cytogenetics. Academic, New York, pp 285–299

Uchida IA, Freeman VCP (1985) Triploidy and chromosomes. Am J Obst Gyn 151: 65–69

Vogel F, Röhrborn G, Schleiermacher E (1971) Chemisch-induzierte Mutationen bei Säuger und Mensch. Naturwissenschaften 58: 131–141

Wagenbichler P, Killian W, Rett A, et al (1976) Origin of the extra chromosome no. 21 in Down's syndrome. Hum Genet 32: 13–16

Webb GC, Krumins EJ, Leversha MA, et al (1984) 49,XXYY,+18 in a liveborn male. J Med Genet 21: 232

Zuffardi O, Tiepolo L (1982) Frequencies and types of chromosome abnormalities associated with human male infertility. In: Crosignani PG, Rubin BL (eds) Genetic control of gamete production and function. Academic, London, pp 261–273

XVII

Human Sex Determination and the Y Chromosome

The Y Chromosome

The human male has two sex chromosomes, one X and one Y. The Y chromosome is in the same size range as the two smallest pairs (the G group) of autosomes. However, the following features enable one to distinguish the Y, even with ordinary staining techniques, from the G chromosomes: the short arm does not have satellites, a constriction is sometimes visible in the middle of the long arm, the chromatids of the long arm stick together, and the distal end of the long arm often looks fuzzy. During the mitotic cycle, the Y replicates later than do the G chromosomes.

The Q-banding pattern of the Y chromosome is much more spectacular than that of the G-group chromosomes, thanks to the very bright distal segment of the long arm (Fig. XVII.1b). Without causing any apparent phenotypic effects, the length of this segment may range from practically zero, which is exceedingly rare, to two or three times its length in the average Y chromosome. Orientals have been reported to have longer Y chromosomes, on the average, than Caucasians (Cohen et al, 1966). The length of the brightly fluorescent distal segment is reflected in the size of a bright Y-body that can be seen in interphase nuclei. A sperm with a Y chromosome is distinguishable from X-carrying sperms by its bright fluorescent spot. In cells with two Y chromosomes, two Y-bodies can be seen (Fig. XVII.1c).

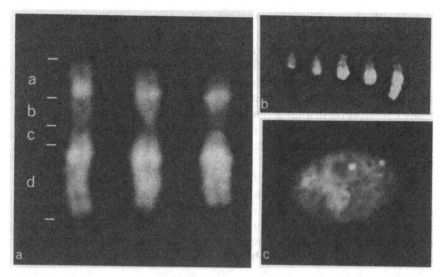

Figure XVII.1. (a) Q-banded human X chromosomes (Therman et al, 1974); (b) variation in the size of the human Y chromosome (Dutrillaux, 1977); (c) two Y bodies in a buccal cell from a 47,XYY male.

Sex Determination in Man

The mechanisms of sex determination may differ greatly, even among closely related groups (cf. Bull, 1983). A sex chromosome mechanism is only one possibility, and even among such mechanisms considerable variation occurs. In the plant *Melandrium*, in which abnormal sex chromosome constitutions have been studied extensively, it is the *presence* of the Y chromosome that primarily determines male development. In *Drosophila*, sex is determined by the *ratio* of X chromosomes to the haploid sets of autosomes. Sex determination in man and in other mammals resembles the plant *Melandrium*, rather than *Drosophila*. The finding that individuals with the chromosome complement 45,X are females and those with 47,XXY males showed that the Y chromosome determined the male sex in man. The basic plan of human sex differentiation is female; the Y chromosome turns an individual into a male. A Y chromosome guarantees a male phenotype (albeit an abnormal one) even if four X chromosomes are present.

That the male-determining factor(s) is situated on the short arm of the Y chromosome was demonstrated by the observation that individuals with an isochromosome for Yq, in addition to an X, are females. Phenotypically they usually are identified as belonging to Turner's syndrome. In other words, they resemble females with a 45,X chromosome constitution (for example, Robinson and Buckton, 1971; Magnelli et al, 1974).

A girl with several Turner symptoms and a deletion of Yp (she had one X chromosome) was described by Rosenfeld et al (1979).

Another type of abnormal Y chromosome, found usually in mosaic individuals with a 45,X cell line, has been described (cf. Cohen et al, 1973). Such chromosomes have two Y long arms, two centromeres, and part of the short arms between them. Individuals with such a chromosome constitution display a phenotype ranging from female to abnormal male, depending on the length of the Yp included and the ratio of the two cell lines. Two Y chromosomes may also be attached long arm to long arm (cf. Daniel et al, 1980).

Dicentric Y chromosomes in which one centromere is inactivated—they resemble an asymmetrical i(Yq)—are found in some mosaics (Schmid and D'Apuzzo, 1978; Taylor et al, 1978). Very small cross-like sex chromosomes, which are difficult to interpret, have also been described, mostly in males. They may represent Yq− chromosomes, iso-chromosomes for Yp, or possibly fragments of X chromosomes.

H-Y Antigen

Sexual differentiation involves several steps, starting with sex determination and continuing with gonadal differentiation. The hormones excreted by the gonadal tissues induce the development of genitals and secondary sex characteristics.

Recent studies have considerably increased our understanding of the primary (gonadal) sex determination. The testis-inducing principle is H-Y antigen, which is a plasma membrane protein. It was originally discovered as a transplantation antigen in that in some inbred mouse strains male skin grafts were rejected by female recipients, but not vice versa (cf. Wachtel, 1983).

The field of gonadal sex determination is under intensive study, and the hypotheses prevailing now are bound to undergo changes when more observations are made. A scheme is presented below which for the present seems most plausible (cf. Wolf, 1981, 1983; Wachtel, 1983).

The structural gene for the H-Y antigen is located on an (unknown) autosome. Factors regulating the expression of this gene are situated both on Yp and the tip of Xp, which remains active on the inactive X chromosome. Other genes known on this segment are the Xg blood group gene and the gene for steroid sulfatase (STS). However, although these genes remain active on the inactive X, they do not show the full activity of the corresponding genes on the active X (cf. Wolf, 1983).

The H-Y antigen regulator on Xp is able to suppress the autosomal gene; as a result, no H-Y antigen is produced in normal females with 46,XX. A male phenotype in 46,XY individuals is induced by one X factor suppressing one autosomal gene, while the other autosomal gene produces the full amount of H-Y antigen. The factor on Yp either suppresses the suppressor on Xp or activates the structural gene. In any case, the Y factor is a powerful one, since individuals with even a 49,XXXXY chromosome constitution are males.

The male-determining effect of H-Y antigen is a threshold phenomenon. Thus, in persons with 45,X, one autosomal gene is suppressed while the other produces H-Y antigen at a level halfway between that of normal males and females, and the result is an abnormal female. Similar low amounts of H-Y antigen have been found in individuals with 46,X,i(Xq) or 46,XXp— in whom one repressor gene has been deleted.

The general rule is that, whenever testicular tissue differentiates, H-Y antigen is present; however, the opposite is not always true. Sex determination in man depends on the interplay of at least four factors: the H-Y structural gene, the suppressor of the structural gene on Xp, the activator gene on Yp, and a factor which determines the development of a gonad-specific receptor. With the exception of spermatogonial cells, H-Y antigen is produced by all male tissues, especially the Sertoli cells in the testis. H-Y antigen is bound by gonad-specific receptors, and this leads to the differentiation of testicular tissue (cf. Müller and Schindler, 1983; Wachtel, 1983).

Genes on the Y Chromosome

Apart from the male-determining factor(s) on the Y chromosome, the human Y chromosome seems to be peculiarly "empty," even considering that the Q-bright distal segment consists of constitutive heterochromatin.

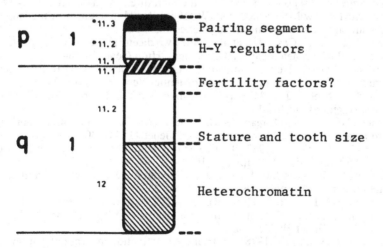

Figure XVII.2. Cytogenetic diagram of the human Y chromosome.

However, factors influencing the tooth size and probably the overall height of the individual seem to be located on the long arm of the Y chromosome (Fig. XVII.2) (Alvesalo, 1978; Alvesalo and de la Chapelle, 1981).

Abnormal Y Chromosomes

Apart from the polymorphism of the bright distal region on Yq, a number of structurally abnormal Y chromosomes have been found in man. The isochromosomes for Yq, Yp, and dicentrics in which the two Y chromosomes are attached, short arm to short arm, have already been described. Dicentrics in which the fusion takes place between two long arms also exist. An inversion that places the centromere at the border of the Q-bright region on the Yq does not seem to have an effect on the phenotype or fertility and has been found to be inherited throughout a very big pedigree (Friedrich and Nielsen, 1974). Several Y ring chromosomes have also been described. Translocation chromosomes in which the Y chromosome is involved are discussed in Chapter XX.

References

Alvesalo L (1978) Tooth sizes in a male with 46,Xdel(Y)(qll) chromosome constitution. IADR Abstr A 57

Alvesalo L, Chapelle A de la (1981) Tooth sizes in two males with deletions of the long arm of the Y-chromosome. Ann Hum Genet 45: 49–54

Bull JJ (1983) Evolution of sex determining mechanisms. Benjamin/Cummings, Menlo Park, California

Cohen MM, MacGillivray MH, Capraro VJ, et al (1973) Human dicentric Y chromosomes: case report and review of the literature. J Med Genet 10: 74–79

Cohen MM, Shaw MW, MacCluer JW (1966) Racial differences in the length of the human Y chromosome. Cytogenetics 5: 34–52

Daniel A, Lyons N, Casey JH, et al (1980) Two dicentric Y isochromosomes, one without the Yqh heterochromatic segment. Hum Genet 54: 31–39

Dutrillaux B (1977) New chromosome techniques. In: Yunis JJ (ed) Molecular structure of human chromosomes. Academic, New York, pp 233–265

Friedrich U, Nielsen J (1974) Pericentric inversion Y in a population of newborn boys. Hereditas 76: 147–152

Magnelli NC, Vianna-Morgante AM, Frota-Pessoa O, et al (1974) Turner's syndrome and 46,X,i(Yq) karyotype. J Med Genet 11: 403–406

Müller U, Schindler H (1983) Testicular differentiation - a developmental cascade. Differentiation 23, Suppl: 99–103

Robinson JA, Buckton KE (1971) Quinacrine fluorescence of variant and abnormal human Y chromosomes. Chromosoma 35: 342–352

Rosenfeld RG, Luzzatti L, Hintz RL, et al (1979) Sexual and somatic determinants of the human Y chromosome: studies in a 46,XYp− phenotypic female. Am J Hum Genet 31: 458–468

Schmid W, D'Apuzzo V (1978) Centromere inactivation in a case of Turner variant with two dicentric iso-long arm Y chromosomes. Hum Genet 41: 217–223

Taylor MC, Gardner HA, Ezrin C (1978) Isochromosome for the long arm of the Y in an infertile male. Hum Genet 40: 227–230

Therman E, Sarto GE, Patau K (1974) Center for Barr body condensation in the proximal part of the human Xq: a hypothesis. Chromosoma 44: 361–366

Wachtel SS (1983) H-Y antigen and the biology of sex determination. Grune and Stratton, New York

Wolf U (1981) Genetic aspects of H-Y antigen. Hum Genet 58:25–28

Wolf U (1983) X-linked genes and gonadal differentiation. Differentiation 23, Suppl: 104–106

XVIII
Human X Chromosome

The Structure of the X Chromosome

The normal female chromosome complement contains, in addition to 44 autosomes, two X chromosomes that are members of the medium-sized C group. The X chromosome constitutes 5.3 percent of the length of the haploid karyotype, and its centromere index is 0.38. The fluorescent pattern of the X chromosome is shown in Fig. XVII.la. In the short arm, two main regions can be seen: a brighter distal segment (a) and a less Q-bright region (b) next to the centromere. On the other side of the centromere is another Q-dark segment (c), which is shorter than the corresponding region on the Xp (b). The rest of the Xq is Q-bright (d) and is divided into two almost equal parts by a narrow darker band.

X Chromatin or Barr Body Formation

In 1949 Barr and Bertram discovered that the nuclei of the nerve cells of the female cat had a condensed, deeply stained body that was absent in the cells of the males. This X chromatin (also called sex chromatin or Barr body) was later found to represent one X chromosome that is con-

→

Figure XVIII.1. Behavior of the inactive X chromosome(s). (a) A drumstick in a polymorphonuclear blood cell; (b) two Barr bodies in the buccal nucleus of a 47,XXX woman; (c) one Barr body in a normal woman; (d) inactive X chromosome in early prophase; (e) Barr body in a fibroblast; (f) inactive X chromosome in late prophase (Feulgen staining, bar = 5μ).

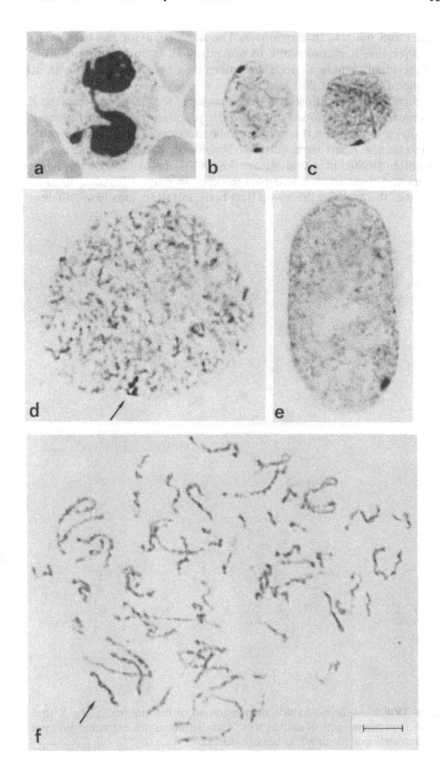

densed and inactive in the female. The same behavior has been established for one X chromosome in practically all mammalian females. If an individual, female or male, has more than two X chromosomes, all but one of them form Barr bodies.

Figure XVIII.1d–f illustrates the behavior of the Barr body-forming X chromosome in interphase and through prophase in human fibroblast cells. Even in prophase it continues to be more condensed than the other chromosomes. Apart from cultured fibroblasts, Barr bodies in man are most often studied in buccal smears (Fig. XVIII.1b–c), but also in vaginal smears and sometimes in hair-root follicles. In normal females, 30 to 50 percent of the buccal cells show a Barr body, but the counts from different

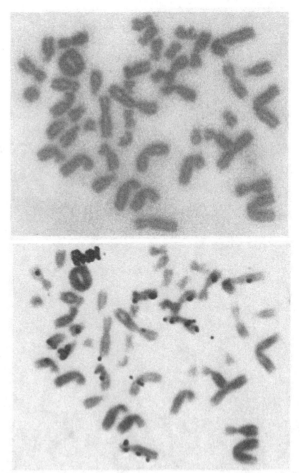

Figure XVIII.2. Autoradiographic demonstration of the late-replicating X in a human lymphocyte with a double r(9). Top: Before autoradiography. Bottom: After autoradiography; one X is heavily labeled.

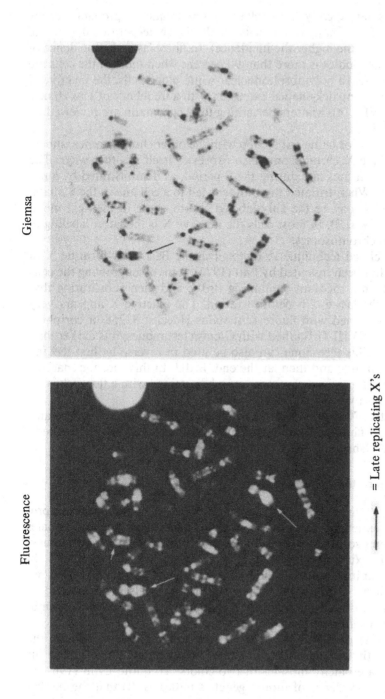

Giemsa

Fluorescence

$\longrightarrow$ = Late replicating X's

$\rightarrow$ = Early replicating X

Figure XVIII.3. Two late-replicating X chromosomes (long arrows) and one early-replicating X (short arrow) in a woman with 47,XXX (BrdU, left stained with Hoechst 33258, right with Giemsa) (Latt et al, 1976).

laboratories vary greatly. This relatively low frequency probably results from the fact that many of the cells are dead (the deeper one digs in the buccal mucosa, the higher the incidence). In fibroblasts, the incidence of cells with Barr bodies is more than 90 percent when most of the cells are in the G_2 phase. In polymorphonuclear white blood cells, the inactive X appears as a drumstick-shaped extrusion with a frequency of 1 to 10 percent (Fig. XVIII.1a). Matters pertaining to X chromatin are reviewed by Mittwoch (1974).

The allocycly, or being out of step with the other chromosomes, shown by the inactive X chromosome also expresses itself in other ways. The inactive X replicates late during the S period, as demonstrated by autoradiography. When tritiated thymidine is fed to a cell late in the S phase, the inactive X may be the only labeled chromosome in the next metaphase (Fig. XVIII.2). In most cells the inactive X is the latest labeling of the C-group chromosomes.

A more refined technique to distinguish the Barr body-forming X in metaphase has been invented by Latt (1973). It involves growing the cells for 40–44 h in a medium containing BrdU, and then substituting thymidine for the last 6–7 h before fixation. The inactive X appears very bright when stained with fluorescent stains Hoechst 33258 or coriphosphine O (Fig. XVIII.3). Stained with Giemsa techniques, it is darker than the other X. This technique can also be used in reverse by first feeding the cells thymidine and then, at the end, BrdU. In this case the inactive X is the least fluorescent chromosome when stained with a fluorochrome and the faintest when stained with Giemsa.

The inactive X is often shorter in metaphase than the active one and may show a differential staining even with ordinary cell culture methods (Takagi and Oshimura, 1973; Sarto et al, 1974).

Inactivation of the X Chromosome

Unlike the Y chromosome, the human X seems to contain genes in proportion to its length. Despite the fact that males have one and females two X chromosomes, the sexes do not differ from each other very much, apart from sexual development and secondary sex characteristics. In many other animal groups, gender differences are considerable. This relative similarity of the male and female in mammals is achieved by a mechanism of dosage compensation that allows all but one X chromosome in each cell to be turned off.

Lyon (1961) proposed the single-active X hypothesis to explain the observation that in the mouse, females heterozygous for X-linked fur color genes are patchy mosaics of two colors. To quote Lyon (1961, p. 372): ". . .the evidence of mouse genetics indicates: (1) that the heteropycnotic X-chromosome can be either paternal or maternal in origin in different cells of the same animal; (2) that it is genetically inactivated."

This hypothesis has inspired an enormous amount of research, and the following principles have been established with respect to X-inactivation.

The inactivation of one X chromosome in the normal female, or of all but one X in individuals with several X chromosomes, takes place early in embryonic development at an estimated 1000- to 2000-cell stage of the blastocyst or possibly even earlier (cf. Lyon, 1974). The inactivation involves the paternal or maternal X chromosome at random, and once an X chromosome is turned off, the event seems to be irreversible. In all the descendants of the cell in which the inactivation took place, the same X is turned off and forms the Barr body. The inactive X is not transcribed; it is facultatively heterochromatic. It has recently been shown that the tip of Xp always remains active (cf. Wolf, 1983); this will be discussed in Chapter XX.

The single-active X hypothesis implies that a female heterozygous for a gene on the X chromosome should have two different cell populations, each with a different allele being expressed. Studies of patchy fur colors, which have yielded so much information in the mouse, are not possible in human populations. However, in a few human conditions, a patchy appearance resulting from X-inactivation seems to be expressed, for instance, in ocular albinism and in choroidemia (cf. Gartler and Andina, 1976).

On the cellular level, this phenomenon can be studied for a few X-linked genes (cf. Migeon, 1979). The first demonstration of the expression of a single X was done by Davidson et al (1963). They cloned fibroblasts from a woman heterozygous for two allozymes of the enzyme G6PD (glucose–6–phosphate dehydrogenase). On starch-gel electrophoresis, the uncloned cells showed the double pattern, whereas clones of cells revealed only one or the other allozyme.

In a mouse with patchy fur color and with a normal and an abnormal X chromosome (so-called Cattanach translocation), one X formed the Barr body in one type of patch, and the other X formed the Barr body in other patches. This observation was in accord with the known fur-color genes on the two X chromosomes (cf. Cattanach, 1975). Another building block in the evidence that the late-replicating X chromosome is also the inactive one was provided by observations on mules and hinnies (reviewed by Gartler and Andina, 1976). In these hybrids the donkey-X and the horse-X can be distinguished morphologically. Each species has its own type of G6PD, and the predominant G6PD isozyme in both fibroblasts and erythrocytes of the two hybrids turned out to be the one characteristic of the horse. In agreement with this observation in the two hybrids, the donkey-X was late-replicating in about 90 percent of the cells and horse-X in 10 percent. Whether this inequality of expression of the two X chromosomes depends on original nonrandomness of the inactivation, or originally random inactivation, followed by selection for the cell line in which the horse-X is active, is not known.

Reactivation of the X Chromosome

As already pointed out, once X inactivation has taken place, it seems to be irreversible. All experimental attempts, whether by treatment with chemicals or hybridization with cells of other mammals, have failed to reactivate the inactive X chromosome. However, a few genes have been found to be derepressed spontaneously (Kahan and DeMars, 1975). Furthermore, treatment with the demethylating agent 5-azacytidine has been observed to derepress some genes (Mohandas et al, 1981; Lester et al, 1982). This has led to the hypothesis that X inactivation on the molecular level involves the methylation of DNA (cf. Gartler and Riggs, 1983; Mohandas and Shapiro, 1983).

The inactive X is reactivated in the oocytes some time before meiosis, the exact timing being still under dispute (cf. Gartler and Andina, 1976). Both X chromosomes are transcribed, and neither shows heteropycnotic behavior. In meiosis the two X chromosomes pair normally and the bivalent formed by them does not differ from autosomal bivalents (Fig. XV.6). Whether the inactive X is spontaneously derepressed under very abnormal conditions, as in malignant tumors, is still unclear (cf. Sandberg, 1983; Takagi et al, 1983).

Interestingly, in male meiosis the opposite phenomenon occurs. Both the X and Y chromosomes appear heteropycnotic from zygotene to diplotene and are not transcribed during spermatogenesis (cf. Cattanach,1975).

It seems to be a general characteristic of the sex chromosomes, at least at certain stages of development, that they are out of step with the autosomes; they have a tendency to allocycly. This is correlated with their inactivity during the same stages.

Sex Reversal and Intersexuality

It is obvious that, apart from the main sex-determining genes, discussed above, many other genes, some of them autosomal, affect sexual development in man. In many animal species, such as *Drosophila*, mouse, rat, and goat, genes are known that can reverse the chromosomal sex. In man, too, there are diverse conditions in which an individual develops a phenotype that is the opposite of his or her chromosomal sex. In most cases the cause of such sex reversal is a mutant gene. The different types of XY women have been reviewed by Sarto (1972).

Individuals with *XY gonadal dysgenesis* (or *Swyer's syndrome*) are women with streak gonads. They fall into two groups. In one type the H-Y antigen is on the male level, but a mutation of an X-linked gene, which is responsible for the gonad-specific receptors, makes these individuals unable to develop testes (cf. Wachtel, 1983). The affected individuals are of greater than average height, and, because of the streak gonads, no secondary sex characteristics develop. The other type of individuals with Swyer's syndrome are similar in phenotype, but they lack the H-Y antigen. Furthermore, there exists an extremely rare gonadal agenesis in women with XY sex chromosome constitution (Sarto and Opitz, 1973). One of the best known of the conditions in which an individual with the XY sex chromosome constitution develops a female phenotype is the *testicular feminization syndrome*. In the rat and the mouse, a corresponding gene (Tfm) is located on the X. In man, too, the Tfm gene has been found to be X-linked (cf. Migeon et al, 1979), and it has been localized to the segment Xql2-Xpll (Migeon et al, 1981).

Families in which the testicular feminization gene is inherited are characterized by a predominance of female births. The affected individuals usually have a fully feminine appearance and often come to the attention of a physician only because of primary amenorrhea (lack of menstruation) and sterility. They have testes in the abdominal cavity or the inguinal canal, but the mutant gene renders the target organs insensitive to androgens produced by the testes (cf. Wachtel, 1983). It is important that these women and others with 46,XY chromosome constitution be diagnosed early enough, because their streak gonads often undergo malignant transformation and should be removed prophylactically.

More than 150 males with a 46,XX (or rarely a 45,X) chromosome constitution have been described. They have male levels of the H-Y antigen and are sterile (cf. de la Chapelle, 1972). They resemble somewhat males with Klinefelter's syndrome (47,XXY), but are shorter. Their tooth size is also in the female range, which is to be expected, since they obviously lack the part of Yq in which the determinants for height and tooth size are assumed to be situated (Alvesalo and de la Chapelle, 1979).

The origin of the XX males is still unclear; more probably than not it is heterogeneous. The claims that a visible part of Yp is translocated to Xp (cf. Wachtel, 1983) have not been confirmed (de la Chapelle et al, 1979). This naturally does not rule out a microscopically undetectable exchange between the short arms of Y and X chromosomes, or a translocation from Yp to an autosome. A number of other explanations have also been put forward (cf. de la Chapelle, 1983; Wachtel, 1983).

A further rare sex reversal syndrome in which individuals with 46,XY chromosomes are females is campomelic dysplasia with an autosomal recessive inheritance (cf. Wolf, 1981). In XX true hermaphroditism the affected individuals have both ovarian and testicular tissue (cf. Wolf, 1981; Wachtel, 1983).

References

Alvesalo L, Chapelle A de la (1979) Permanent tooth sizes in 46,XX-males. Ann Hum Genet 43: 97–102

Barr ML, Bertram EG (1949) A morphological distinction between neurones of the male and female, and the behavior of the nucleolar satellites during accelerated nucleoprotein synthesis. Nature 163: 676–677

Cattanach BM (1975) Control of chromosome inactivation. Annu Rev Genet 9: 1–18

Chapelle A de la (1972) Analytic review: nature and origin of males with XX sex chromosomes. Am J Hum Genet 24: 71–105

Chapelle A de la (1983) The origin of XX males. In: Sandberg AA (ed) Cytogenetics of the mammalian X chromosome, Part B. Liss, New York, pp. 75–85.

Chapelle A de la, Simola K, Simola P, et al (1979) Heteromorphic X chromosomes in 46,XX males? Hum Genet 52: 157–167

Davidson RG, Nitowsky HM, Childs B (1963) Demonstration of two populations of cells in the human female heterozygous for glucose–6-phosphate dehydrogenase variants. Proc Natl Acad Sci USA 50: 481–485

Gartler SM, Andina RJ (1976) Mammalian X-chromosome inactivation. In: Harris H, Hirschhorn K (eds) Advances in human genetics, Vol 7. Plenum, New York

Gartler SM, Riggs AD (1983) Mammalian X-chromosome inactivation. Annu Rev Genet 17: 155–190

Kahan B, DeMars RI (1975) Localized derepression on the human inactive X chromosome in mouse-human cell hybrids. Proc Natl Acad Sci USA 72: 1510–1514

Latt SA (1973) Microfluorometric detection of deoxyribonucleic acid replication in human metaphase chromosomes. Proc Natl Acad Sci USA 70: 3395–3399

Lester SC, Korn NJ, DeMars R (1982) Derepression of genes on the human inactive X chromosome: evidence for differences in locus-specific rates of derepression and rates of transfer of active and inactive genes after DNA-mediated transformation. Somatic Cell Genet 8: 265–284

Lyon MF (1961) Gene action in the X-chromosome of the mouse (*Mus musculus* L.). Nature 190: 372–373

Lyon MF (1974) Mechanisms and evolutionary origins of variable X-chromosome activity in mammals. Proc R Soc Lond B 187: 243–268

Migeon BR (1979) X-chromosome inactivation as a determinant of female phenotype. In: Vallet HL, Porter IH (eds) Genetic mechanisms of sexual development. Academic, New York, pp 293–303

Migeon BR, Brown TR, Axelman J, et al (1981) Studies of the locus for androgen receptor: Localization on the human X chromosome and evidence for homology with the *Tfm* locus in the mouse. Proc Natl Acad Sci USA 78: 6339–6343

Migeon CJ, Amrhein JA, Keenan BS, et al (1979) The syndrome of androgen insensitivity in man: its relation to our understanding of male sex differentiation. In: Vallet HL, Porter IH (eds) Genetic mechanisms of sexual development. Academic, New York, pp 93–128

Mittwoch U (1974) Sex chromatin bodies. In: Yunis JJ (ed) Human chromosome methodology, 2nd edn. Academic, New York, pp 73–93

Mohandas T, Shapiro LJ (1983) Factors involved in X-chromosome inactivation. In: Sandberg AA (ed) Cytogenetics of the mammalian X chromosome, Part A. Liss, New York, pp 271–297

Mohandas T, Sparkes RS, Shapiro LJ (1981) Reactivation of an inactive human X-chromosome: evidence for X inactivation by DNA methylation. Science 211: 393–396

Sandberg AA (1983) The X chromosome in human neoplasia, including sex chromatin and congenital conditions with X-chromosome anomalies. In: Sandberg AA (ed) Cytogenetics of the mammalian X chromosome, Part B. Liss, New York, pp 459–498

Sarto GE (1972) Genetic disorders affecting genital development in 46,XY individuals (male pseudohermaphroditism). Clin Obstet Gynecol 15: 183–202

Sarto GE, Opitz JM (1973) The XY gonadal agenesis syndrome. J Med Genet 10: 288–293

Sarto GE, Therman E, Patau K (1974) Increased Q fluorescence of an inactive Xq— chromosome in man. Clin Genet 6: 289–293

Takagi N, Oshimura M (1973) Fluorescence and Giemsa banding studies in the allocyclic X chromosome in embryonic and adult mouse cells. Exp Cell Res 78: 127–135

Takagi N, Yoshida MA, Sugawara O, et al (1983) Reversal of X-inactivation in female mouse somatic cells hybridized with murine teratocarcinoma stem cells in vitro. Cell 34: 1053–1062

Wachtel SS (1983) H-Y antigen and the biology of sex determination. Grune and Stratton, New York

Wolf U (1981) Genetic aspects of H-Y antigen. Hum Genet 58: 25–28

Wolf U (1983) X-linked genes and gonadal differentiation. Differentiation 23, Suppl: 104–106

XIX
Numerical Sex Chromosome Abnormalities

Aneuploidy of X Chromosomes in Individuals with a Female Phenotype

The sex chromosomes show a much wider range of viable aneuploidy than do the autosomes, presumably for the following reasons. On the one hand, the Y chromosome seems to contain very few genes apart from those determining the male sex; on the other, all but one X chromosome in a cell are inactivated, forming X chromatin bodies in the interphase. This rule can be stated another way: there is one active X chromosome for each diploid complement of autosomes.

Figure XIX.1 summarizes the nonmosaic numerical sex chromosome abnormalities found so far, their incidence in newborn infants of the same sex, and the number of Barr bodies formed by them. One male with XYYYY has also been described (Sirota et al, 1981).

Aneuploidy results from nondisjunction in either the meiotic division of the parents or the early cleavage divisions of the affected individuals. Aneuploidy for more than one chromosome is the result of nondisjunction in both meiotic divisions, nondisjunction in the meiosis of both parents (which must be a rare coincidence), or abnormalities in more than one mitosis.

The incidence of 47,XXX and 47,XXY children increases with maternal age, as does that of autosomal trisomies. In these cases nondisjunction apparently occurs mainly in maternal meiosis. The incidence of 45,X children, on the other hand, seems to be independent of maternal age. Indeed, studies of the Xg blood group gene, which is located on the X chromosome, show that in 78 percent of the cases, the gamete without a sex chromosome came from the father (Sanger et al, 1971). The syn-

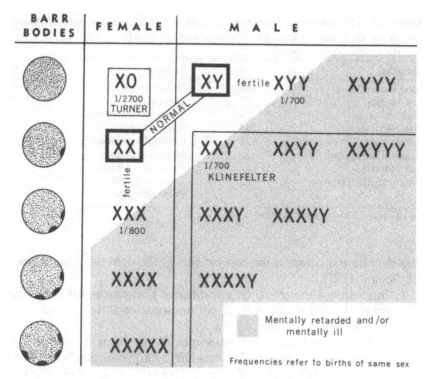

Figure XIX.1. The human X and Y chromosome constitutions and the Barr bodies formed by the X chromosomes. (The male with 49,XYYYY is missing from the diagram.)

dromes caused by the various sex chromosome aneuploidies have been reviewed by Hamerton (1971) and de Grouchy (1974).

Individuals with a 45,X chromosome constitution develop the so-called Turner's syndrome. They are phenotypic females, but their gonads, which appear normal in the embryo, degenerate postnatally and later consist mainly of fibrous tissue that lacks follicles. Such ovaries do not produce estrogens, and the patients remain sexually infantile. A large number of phenotypic abnormalities have been described in these patients, the most common one being short stature (under 5 ft or 150 cm). Other anomalies (Table XIX.l), such as webbed neck, low hairline, shield chest, cubitus valgus (changed carrying angle of the elbow), and pigmented nevi are somewhat less frequent. Often the cardiovascular and urinary systems are also affected.

Patients with Turner's syndrome do not seem to be severely mentally retarded any more often than females with two X chromosomes. Intelligence tests show that their verbal IQs are equal to normal controls, but that their performance IQs are lower. This has been interpreted to mean

Table XIX.1. Frequency of Turner Symptoms in 332 Individuals with 45,X in Percentage of Those Patients for Whom Symptom Was Recorded. All Ascertainment Types Pooled.[a]

Symptom	%	Symptom	%
Short stature	100	Webbed neck	42
Low hairline	72	Short neck	77
Shield chest	74	Urinary system anomaly	44
Pigmented nevi	64	Hypertension	37
Cubitus valgus	77	Gonadal dysgenesis	91
Short 4th metacarpal	55	Mental retardation	19
Nail anomaly	57	Retarded bone age	64
Cardiovascular anomaly	23	Thyroid disease	18
		Color blindness	11

[a]Denniston and Ulber, unpublished data.

that they have a defect in the cortical part of the right brain hemisphere (cf. Polani, 1977).

An interesting although so far unexplained phenomenon is that about 97 percent of zygotes with a 45,X chromosome constitution end up as spontaneous abortions (Carr, 1971).

Individuals with three X chromosomes do not seem to form a well-defined syndrome. They are often mentally retarded or psychotic. Their mental status, as well as that of persons with other types of sex chromosome anomalies, is reviewed by Polani (1977). Triple-X women, although fertile, produce overwhelmingly normal children instead of one-half of the daughters with 47,XXX and one-half of the sons with 47,XXY chromosome constitutions, as might be expected. One possible explanation is that in the first meiotic division of the oocyte, the extra chromosome always ends up in the polar body. Other organisms in which aberrant chromosomes are relegated to the polar bodies are found in *Drosophila* and moths.

Patients with more than three X chromosomes suffer from severe mental retardation and have several somatic anomalies; however, their sex development is usually normal. Nielsen et al (1977) reviewed the 26 cases of 48,XXXX found up to that time. They showed a wide range of abnormalities but varied considerably from case to case. The total of only 26 patients found up to this time with this type of aneuploidy shows how rare these multiple sex chromosome anomalies actually are.

As Fig. XIX.1 demonstrates, more than one X chromosome can be found in individuals with a male phenotype. They are discussed in the following paragraphs.

Sex chromosome aneuploidies—mainly XO, XXY, and XXX—are also described in a number of other mammals (cf. Lyon, 1974; Wurster-Hill et al, 1983).

Sex Chromosome Aneuploidy with a Male Phenotype

In the group of sex chromosome aneuploidies with a male phenotype, the XXY and XYY conditions are about equally frequent at birth. Individuals with 47,XXY chromosome constitution form a fairly well-defined, so-called Klinefelter's syndrome. They are characterized by a eunuchoid habitus, their small testes are devoid of sperm cells, and they are sterile. They also show a tendency to breast development (gynecomastia). The XXY sex chromosome constitution seems to lower the IQ to some extent, and as a result the affected individuals are often mildly mentally retarded (cf. Murken, 1973). They also show psychotic tendencies and may eventually end up in correctional institutions (cf. Polani, 1977).

Patients with one Y chromosome and more than two X chromosomes are mentally retarded and display various other symptoms. To take one example, males with the 49,XXXXY chromosome constitution have an IQ in the range of 20 to 50, extensive skeletal anomalies, severe hypogenitalism, strabismus, wide-set eyes, and other anomalies.

The 47,XYY syndrome has unfortunately been sensationalized by the news media, which have often depicted these individuals as a group of murderers and other violent criminals. This is not true. What *is* true is that persons with this chromosome complement have a higher probability of coming into conflict with the law than normal males, but their crimes are usually nonviolent (cf. reviews by Hook 1973; Polani, 1977; Witkin et al, 1976). Most males with an XYY sex chromosome constitution lead normal lives and are not distinguishable from normal males except that they are usually considerably taller than their male relatives, often being 6 ft tall and over (>180 cm) (cf. Murken, 1973).

The highest incidence of XYY males is found when prisoners who are 6ft tall or taller are chosen for chromosome studies. Apart from their above- average height, XYY males do not represent a defined syndrome, although some neurological symptoms and specific features of body build have been described in them (Daly, 1969). Apart from some individuals, their fertility does not seem to be impaired to any great extent. The XYY condition is often accompanied by mental retardation (cf. Murken, 1973).

Males with the chromosome constitution 48,XXYY are also found in increased numbers in prisons. They show features in common with Klinefelter's syndrome, are tall and are more or less mentally retarded.

Higher aneuploidies like XXXYY and XYYY are rare, and one individual with XYYYY has also been described (Sirota et al, 1981). Patients with such sex chromosome constitutions are retarded and display numerous anomalies.

Mosaicism

Mosaicism for sex chromosome aneuploidy seems to be considerably more common than for autosomal aneuploidy. Mosaics with either a 46,XX or 46,XY cell line accompanied by a 45,X cell line are the most common, but three or even more different cell lines have been described. Sometimes the type of mosaicism unequivocally reveals its origin. Thus mosaicism of the type 45,X/46,XX/47,XXX obviously arose by nondisjunction in a chromosomally normal female. In more complicated mosaics—and they can be very complicated, indeed—the mechanisms that generated them remain a matter for speculation.

A wide range of conditions representing intersexes in whom the external genitalia are ambiguous, or hermaphrodites who have gonads containing both testicular and ovarian tissue, has been described (review in Hamerton, 1971; Wachtel, 1983). Such persons are often mosaics with the chromosome constitution 45,X/46,XY or 46,XX/46,XY. Whether the individual is more male or more female depends on which cell line is predominant in the different organs. More complicated types of mosaicism are also found in such individuals.

Low-grade mosaicism does not affect the phenotype. Cell lines with 45,X are found in both normal males and females. When the proportion of 45,X cells increases, symptoms of Turner's syndrome begin to appear, and we find in this group all the intermediates from normal female phenotype to full Turner's syndrome. Mitotic nondisjunction, as well as other chromosome aberrations, increases in older persons, who show significantly more aberrant cells than do younger ones. The proportion of 45,X cells increases especially in older women (cf. Galloway and Buckton, 1978).

In mosaic individuals the same rules apply as in nonmosaics, namely, in each cell all X chromosomes but one are inactivated. If the mosaicism arises after the X inactivation has taken place, only cells in which nondisjunction involves the inactive X chromosome(s) seem to be viable (Daly et al, 1977).

References

Carr DH (1971) Chromosomes and abortion. In: Harris H, Hirschhorn K (eds) Advances in human genetics, Vol 2. Plenum, New York, pp 201–257

Daly RF (1969) Neurological abnormalities in XYY males. Nature 221: 472–473

Daly RF, Patau K, Therman E, et al (1977) Structure and Barr body formation of an Xp+ chromosome with two inactivation centers. Am J Hum Genet 29: 83–93

Galloway SM, Buckton KE (1978) Aneuploidy and ageing: chromosome studies on a random sample of the population using G-banding. Cytogenet Cell Genet 20: 78–95

Grouchy J de (1974) Sex chromosome disorders. In: Busch H (ed) The cell nucleus, Vol II. Academic, New York, pp 415–436

Hamerton JL (1971) Human cytogenetics. Clinical cytogenetics, Vol II. Academic, New York

Hook EB (1973) Behavioral implications of the human XYY genotype. Science 179: 139–150

Lyon MF (1974) Mechanisms and evolutionary origins of variable X-chromosome activity in mammals. Proc R Soc Lond B 187: 243–268

Murken J-D (1973) The XYY-syndrome and Klinefelter's syndrome. George Thieme, Stuttgart

Nielsen J, Homma A, Christiansen F, et al (1977) Women with tetra-X (48,XXXX). Hereditas 85: 151–156

Polani PE (1977) Abnormal sex chromosomes, behavior and mental disorder. In: Tanner JM (ed) Developments in psychiatric research. Hodder and Stoughton, London, pp 89–128

Sanger R, Tippett P, Gavin J (1971) Xg groups and sex abnormalities in people of northern European ancestry. J Med Genet 8: 417–426

Sirota L, Zlotogora Y, Shabtai F, et al (1981) 49,XYYYY. A case report. Clin Genet 19: 87–93

Wachtel SS (1983) H-Y antigen and the biology of sex determination. Grune and Stratton, New York

Witkin HA, Mednick SA, Schulsinger F, et al (1976) Criminality in XYY and XXY men. Science 193: 547–555

Wurster-Hill DH, Benirschke K, Chapman DI (1983) Abnormalities of the X chromosome in mammals. In: Sandberg AA (ed) Cytogenetics of the mammalian X chromosome, Part B. Liss, New York, pp 283–300

XX
Structurally Abnormal X Chromosomes

Abnormal X Chromosomes Consisting of X Material

The wide range of aneuploidy displayed by the human X chromosome is matched by a greater variety of structural abnormalities than is found in any of the autosomes. In cells with one normal and one abnormal X chromosome, consisting of X chromosomal material, the abnormal X is practically always inactivated. This naturally diminishes the effect of the abnormal X. The sizes of the Barr bodies reflect the sizes of the abnormal X chromosomes. They range from considerably smaller than normal for partly deleted chromosomes to very large ones for chromosomes consisting of two long arms of the X or even of two almost intact X chromosomes.

Figure XX.1 illustrates, apart from X-autosomal or X-Y translocation chromosomes, the main types of abnormal X chromosomes found in man (Therman et al, 1974). The most common is the isochromosome for the long arm, i(Xq). An isochromosome, in which by definition the two arms are genetically identical, may result through different mechanisms. The product of a reciprocal translocation between two X chromosomes is not a true isochromosome, but it cannot be distinguished from one by cytological means. A true but dicentric isochromosome is formed when the two sister chromatids of Xp break near the centromere and join. However, the most common mode of origin is probably misdivision of the centromere (Fig. XVI.2), which may also lead to telocentrics of the long arm, tel(Xq) (Therman et al, 1981). Segregation of an adjacent chromatid exchange at identical points between two X chromosomes would also give rise to a cell with i(Xq) and another with 45,X chromosomes,

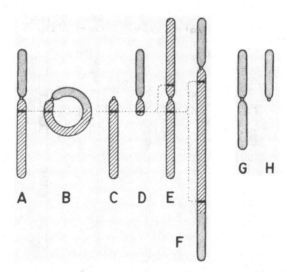

Figure XX.1. Five structurally different X chromosomes found in humans (A–E) and two never observed (G–H). Crossline marks, presumed inactivation center. A, normal X; B, ring X; C, telocentric Xq; D, Xq−; E, isochromosome Xq; F, X;X translocation chromosome: attachment, long arm to long arm; G, isochromosome Xp; H, telocentric Xp (Therman et al, 1974).

and would explain the mosaicism so often found in women with an i(Xq) chromosome.

Deletions of the short or the long arm of the human X chromosome also occur. A deletion of Xp may involve part or all of the arm. In the latter case an acrocentric or telocentric for the long arm is formed. The chromosomes in which the long arm has undergone a deletion range from those resembling a chromosome 18 to metacentrics and to those with even shorter deletions. Short deletions may be difficult to detect cytologically, and they would also cause very few, if any, symptoms (Trunca et al, 1984). For the formation of ring chromosomes a distal segment must be missing from both arms to allow the "sticky" ends to join. X chromosome deletions have been reviewed by Wyss et al (1982) and Maraschio and Fraccaro (1983).

Individuals bearing an X chromosome with a deletion are less affected than are those in whom the whole X is missing. Figure XX.3 gives a rough summary of the symptoms caused by partial deletions of one X chromosome. A deletion of Xp21→pter gives rise to short stature. An interstitial deletion has few if any effects (Herva et al, 1979). The loss of the whole Xp results in a full-blown Turner's syndrome, although primary amenorrhea is considerably rarer than in women with 45,X. Deletions with breakpoints spread over Xq (the c region is never deleted) cause primary or secondary amenorrhea and in some cases a few Turner

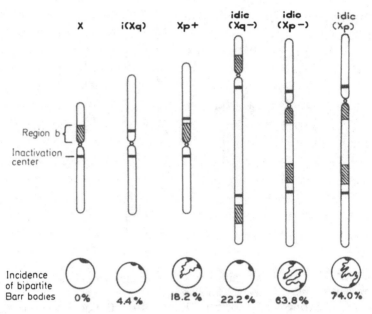

Figure XX.2. The relative positions of the X inactivation center and the presumably always-active region on Xp in six structurally different human X chromosomes, and the incidence and hypothetical structure of bipartite Barr bodies formed by them (in idic(Xp) the attachment is telomeric) (Therman et al, 1980).

symptoms, which possibly may depend on a hidden 45,X cell line. Deletions with breakpoints distal to Xq27 have not been found, possibly because they do not show any symptoms (and are naturally cytologically difficult to recognize) (cf. Maraschio and Fraccaro, 1983).

In a number of patients ascertained for amenorrhea and often showing few other symptoms, a very long chromosome is found in place of one X chromosome. They consist of two X chromosomes attached, short arm to short arm or long arm to long arm, with more or less material missing from the two chromosomes. The second centromere in such chromosomes is inactivated, showing no primary constriction. However, a C-band is visible at its site (Fig. XX.4c). The symptoms shown by the carriers correspond to those caused by Xp and Xq deletions, the women having idic(Xp−) chromosomes exhibiting more abnormalities (cf. Mattei et al, 1981b; Zakharov and Baranovskaya, 1983).

An abnormal X chromosome is often encountered in mosaics in which a second cell line has the chromosome constitution 45,X. In some cases the latter cell line may have come about by the loss of the abnormal X chromosome. In others, both cell types have probably resulted simultaneously from mitotic segregation in a chromatid translocation (Therman

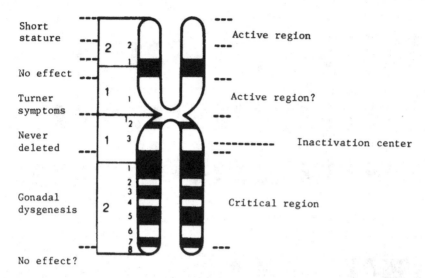

Figure XX.3. Cytogenetic diagram of the human X chromosome showing the presumably always-active regions, the inactivation center, and the critical region. Also indicated at the left are the main phenotypic effects of various deletions.

and Patau, 1974); this explanation has generally been neglected in the literature.

In cells with one normal X chromosome and an abnormal one consisting of X material, the abnormal X forms the Barr body. However, sometimes a few cells are found with the opposite inactivation pattern (Latt, 1979), which speaks for the idea that the two X chromosomes are originally inactivated at random, but that the genetically more balanced cell line gradually takes over (cf. Therman and Patau, 1974).

X Inactivation Center

Whatever the mechanism giving rise to i(Xq) and tel(Xq) chromosomes—and especially misdivision—short-arm isochromosomes and telocentrics should arise with about the same frequency. However, not one verified case of either has been found. If we compare the relatively frequent abnormal X chromosomes with those that probably do not exist (Fig. XX.1), it becomes clear that what the former possess and the latter lack is the Q-dark region next to the centromere on the Xq (c region) (Therman et al, 1974; Therman et al, 1980; Therman and Sarto, 1983).

In persons with i(Xq) chromosomes and those X;X translocation chromosomes that have this region in duplicate, a certain proportion of Barr bodies is bipartite. From these observations the conclusion was drawn that the c-region (Figs. XX.2 and XX.3) contains the *X inactivation cen-*

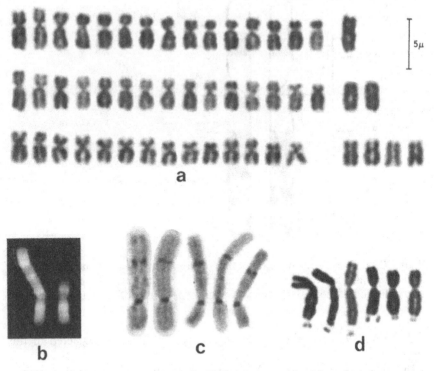

a

b   c   d

Figure XX.4. (a) The C-group and the X chromosome(s) with an inactive centromere from three cells; (b) idic(Xp−) with telomeric attachment and the normal X chromosome; (c) C-banded idic(Xp−) chromosome showing one active and one inactive centromere from five cells; (d) X chromosome expressing the fragile region from six cells ((d) courtesy of S Roberts).

ter, without which the chromosome cannot form a Barr body, and thus remains active. Two active X chromosomes in the same cell would presumably be a nonviable condition. If two inactivation centers are located on the same abnormal X chromosome, each acts as a center for X chromatin condensation, which leads to the formation of bipartite bodies (Therman et al, 1974; Therman and Sarto, 1983).

In a comparison of an X chromosome with the longest possible deletion with a translocation chromosome that contained the shortest possible segment of Xq, with both capable of inactivation, the position of the inactivation center was estimated to be around Xq13 (Therman et al, 1979).

Also arguing for the existence and location of the inactivation center is the fact that those women who are carriers of an unbalanced X;autosome translocation have always been found to have the translocation chromosome with the c region included (Mattei et al, 1981b).

X;Autosome Translocations and the Critical Region

Over 80 balanced X;autosome translocations have been described so far (cf. Wyss et al, 1982; Carpenter, 1983a). All autosomes have been involved, but one-third of the translocations have been between the X and chromosomes 21 and 22. In practically all women with a balanced X;autosome translocation, the normal X is inactivated in the lymphocytes, or in any case the predominant cell line consists of such cells. However, the use of the BrdU technique and the study of other tissues besides lymphocytes have shown that in many cases a minority cell line exists with one of the translocation chromosomes inactivated (cf. Zuffardi, 1983). An interesting example has been presented by Hellkuhl et al (1982). The normal X was inactivated in all lymphocytes of a woman with 46,X,t(X;3)(q28;q21). However, in 117 fibroblasts, the normal X was late-replicating, whereas in 61 cells the Xq+ chromosome was inactivated; the inactivation in the latter type of cell had not spread to the chromosome 3 part.

The balanced carriers fall into two groups in regard to sexual development: normal women, often of proven fertility, and another group characterized by various degrees of gonadal dysgenesis, which in most cases results in primary amenorrhea.

On the basis of six cases of balanced X;autosome translocations, then available, Sarto et al (1973) proposed the hypothesis of the "critical region" on the Xq: if a break is within this region, the carriers suffer from gonadal dysgenesis, whereas a break outside this region does not affect the sexual development. Later Summitt et al (1978) defined the critical region to be Xq13-q27 (Fig. XX.3). Of the cases known to Therman (1983) in the 42 women in whom the break was outside this region, sex development was normal, whereas, of the 29 with the break within the critical region, 22 had primary amenorrhea, 3 secondary amenorrhea, 2 oligomenorrhea, and only 2 were normal women. It is possible that in the Xq deletion cases—as well as in a few inversions and insertions—in which the break falls into the critical region, gonadal dysgenesis is also caused by the same position effect (Therman, 1983).

Balanced X;autosome translocations have been found in 15 males (cf. Carpenter, 1983a). Of the adults, 7/8 were sterile because of azotospermia or oligospermia (lack or scarcity of sperms). Possible explanations for the sterility are: interference with the pairing of X and Y chromosomes, or interference with the inactivation of the X and Y chromosomes in male meiosis.

Persons with an unbalanced X;autosome translocation are often offspring of balanced carrier women. When they are female they have a normal X and a translocation chromosome that consists of parts of an X and an autosome. The translocation chromosome is usually inactivated, and the inactivation may or may not spread to the autosomal part. It is of interest that, whereas in a balanced translocation carrier mother, the normal X is inactivated, in an unbalanced daughter the translocation chromosome forms the Barr body, which shows that the genetically better balanced cell line is taking over. The symptoms of the unbalanced individuals (practically all of them are mentally retarded, with multiple anomalies) depend on which autosomal segment is trisomic, and on the inactivation pattern of the chromosome. In carriers of unbalanced X;autosome translocations the inactivation pattern may vary in different cells.

All these observations agree with the hypothesis that the two X chromosomes in a female, normal or abnormal, are originally inactivated at random and that selection determines the final predominant inactivation pattern (cf. Therman and Patau, 1974).

Always-Active Regions on the Inactive X Chromosome

As mentioned above, we assumed on cytological grounds that the b region always stays active on the inactive X chromosome (Daly et al, 1977; Therman et al, 1976). Since then, genetic evidence has accumulated to prove that the tip of Xp stays active on the inactive chromosome. However, the activity of the genes on it (especially that of STS) is less than the activity of corresponding genes on the active X chromosome (cf. Wolf, 1983).

Since late replication is taken as a sign of inactivation, Otto and Therman (1982) raised the question why early replication should not be considered as reflecting activity. The following regions of the X chromosome are early replicating: the tip of the Xp, and the b and c regions (Fig. XVII.1a). Furthermore, Schempp and Meer (1983) have shown with prophase banding that the sequence of early replication is similar in the tip of Xp and the c region, which contains the inactivation center. It is also possible that the b region behaves in the same way.

Thus, on present evidence the tip of Xp definitely stays active, and the b and c regions may do so. The presence of such regions of the inactive X would be one explanation for the fact that persons with 45,X (or with any of the active regions deleted) or with more than two X chromosomes show phenotypic abnormalities.

Translocations Involving the Y Chromosome

Twenty-four X;Y translocations have been described so far (cf. Wahlström, 1983; Fryns and Van den Berghe, 1983). In all but one, part of the Yq (breakpoint qll) was attached to Xp (breakpoint p22). Both male and female carriers of X;Y translocations have been found (Fryns and Van den Berghe, 1983).

Narahara et al (1978) list 32 Y-autosomal translocation cases. However, those involving the short arms of chromosomes 15 or 22 may not belong to this group, since the satellites and short arms of these chromosomes may show a bright fluorescence, and this bright material may be duplicated and not come from a translocated Y chromosome (Spowart, 1979).

Abnormal X Chromosome Constitutions and the Female Phenotype

The abnormal phenotypes resulting from the various X chromosome abnormalities (Fig. XX.3) raise the question why the carriers should differ from normal women when one and only one X chromosome is active in every diploid cell. Therman et al (1980) and Therman (1983) have reviewed the following hypotheses, which are not mutually exclusive. Indeed, the effects of a specific X chromosome constitution may simultaneously be exerted along different pathways.

1. The quantity of X heterochromatin as such has been assumed to have an effect on the phenotype.
2. An assumed change in the duration of the cell cycle by abnormal X chromosome constitutions has been postulated to have phenotypic effects.
3. A recessive gene(s) in a hemizygous condition might be expressed in the cases in which the same X chromosome is active in all cells.
4. The damage done by abnormal X chromosome constitutions might occur at an early embryonic stage *before* X inactivation.
5. Abnormal X chromosome constitutions could have an adverse effect on ovarian development after the inactive X chromosome(s) has been derepressed in the oocyte.
6. A change in the number of known or presumed always-active regions on the inactive X chromosome may affect the phenotype.
7. The critical region Xq13-q27 has to be intact on both X chromosomes to allow normal female sexual development.
8. Abnormal X chromosome constitutions may exert their effect during the selection period by those cells whose inactivation pattern renders them genetically imbalanced.
9. Irregular or asynchronous inactivation of X chromosomes, especially in the cases with more than two X chromosomes, might affect the phenotype.

The first four hypotheses can, at least for the time being, be discarded for lack of evidence or because they lead to absurd conclusions (Therman, 1983). The position effect of a break in the critical region is based on solid evidence. The remaining hypotheses (effect of the reactivated X chromosomes on oocytes, change in the number of active regions on the inactive X chromosome, effects of cells with a "wrong" inactivation pattern during the selection period, possible irregular or asynchronous inactivation of X chromosomes) probably suffice to explain the various phenotypic effects of the various X chromosome constitutions. I leave it to the reader to try out which hypotheses would best apply to the effects of specific chromosome abnormalities.

Various Behavioral Abnormalities of the X Chromosome

In Fig. XX.4 are collected examples of abnormal behavior characteristic of, but not exclusive to, the X chromosome. Cells with the 45,X chromosome constitution are often significantly increased in older women. At the same time cells that display what at first sight simulates acentric fragments make their appearance. Banding studies have shown these to be X chromosomes in which the centromere is inactivated but shows a C-band (what this involves at the electron microscopic level is not known). Their appearance and behavior are more in agreement with the idea that the centromere is not functioning than with the claim that it has undergone premature division, as the phenomenon has also been interpreted (chromosomes with a prematurely divided centromere do not resemble acentric fragments) (Fitzgerald, 1983). Chromosomes with an inactivated centromere replicate and divide, but their segregation is random and results in some cells lacking these chromosomes while others may have accumulated several of them (Drets and Therman, 1983).

A similar inactivation of the centromere often takes place in dicentric chromosomes, but it seems to be especially common in dicentrics consisting of two X chromosomes, possibly because trisomy for the X is much better tolerated than imbalance for the autosomes. The two X chromosomes are attached either by their short, idic(Xp−), or long arms, idic(Xq−). The site of the inactivated centromere is marked by a C-band, but no Cd band, and the primary constriction is lacking (Fig. XX.4b,c).

Lists of t(X;X) chromosomes are found in Mattei et al (1981b) and in Zakharov and Baranovskaya (1983).

Figure XX.4d illustrates six X chromosomes in which the so-called fragile region is expressed. It appears as a gap at Xq27, and the chromosome is apt to break or form a triradial at this point. As discussed in Chapter VIII, fragile regions are found in several autosomes also. However, the fra(X) is different from the autosomes in that, in the hemizygous state, it has severe phenotypic effects. Most male carriers are mentally retarded (one reason for the higher incidence of mental retardation in men than women). They also display the following somatic anomalies: macrorchidism (large testes), high forehead, prominent nose, and large and protruding ears (cf. Carpenter, 1983b). The fra(X) is inherited in mendelian fashion, and most heterozygous women are normal, whereas the overwhelming majority of male carriers are affected. However, some female carriers are slightly retarded, and even severely retarded female carriers have been described (cf. for example Mattei et al, 1981a).

The fragile region is never expressed in all cells of a carrier. However, its expression can be increased with various modifications of the medium (cf. Glover, 1983; Howard-Peebles, 1983). Whether it is the fragile region as such or a closely linked gene that causes the syndrome is not known. Equally unclear are the pathways from the causative factor to the abnormal phenotype.

Summary of Mammalian X Inactivation

1. Normally each diploid cell has one active X chromosome, the other(s) is inactivated. However, there are exceptions to this rule, especially in X;autosome translocation cases.
2. Inactivation takes place at the blastocyst stage and is random for the paternal and maternal X chromosomes. However, in some tissues, especially the extra-embryonic membranes, the paternal X is preferably inactivated. (In marsupials the paternal X is always inactivated.)
3. No exceptions seem to exist to the rule that inactivation is permanent in all descendants of a cell in somatic tissues: they form a clone.
4. The tip of Xp on the inactive X is always active; however, the level of activity of the genes in this region is less than that of the corresponding genes on the active X. One or two further regions on the inactive X (the Q-dark regions on both sides of the centromere) possibly behave in the same way.
5. Attempts to derepress the inactive X experimentally have not been successful (whether this happens spontaneously in some malignant tumors is still unclear). However, a few genes have been derepressed both spontaneously and experimentally.
6. No exceptions seem to exist to the rule that both (all) X chromosomes are active during meiosis in the oocytes.
7. The inactive X chromosome forms a condensed Barr body and is late-replicating during the S period. Whether inactivation is correlated with these two parameters under all conditions is not clear.
8. When a cell has one normal and one abnormal X consisting of X material, the abnormal chromosome is inactivated (provided it contains the inactivation center). In some cases in a few cells the opposite situation seems to prevail.
9. In carriers of balanced reciprocal X;autosome translocations, the normal X is inactivated. Sometimes in a minority cell line, one of the translocation chromosomes forms the Barr body.
10. In females with an unbalanced X;autosome translocation, the translocation chromosome is inactivated. Inactivation may or may not spread to the autosomal segment. In some patients additional cell lines with different inactivation patterns have been found.

11. The X and Y chromosomes are without exception inactivated in spermatocytes, and apparently stay inactive through meiosis. One explanation for the sterility of males with an X;autosome translocation is that such a chromosome may be unable to undergo inactivation in male meiosis.

References

Carpenter NJ (1983a) Balanced X;autosome translocations and gonadal dysfunction in females and males. In: Sandberg AA (ed) Cytogenetics of the mammalian X chromosome, Part B. Liss, New York, pp 211-224

Carpenter NJ (1983b) The fragile X chromosome and its clinical manifestations. In: Sandberg AA (ed) Cytogenetics of the mammalian X chromosome, Part B. Liss, New York, pp 399-414.

Daly RF, Patau K, Therman E, et al (1977) Structure and Barr body formation of an Xp+ chromosome with two inactivation centers. Am J Hum Genet 29: 83-93

Drets ME, Therman E (1983) Human telomeric 6;19 translocation chromosome with a tendency to break at the fusion point. Chromosoma 88: 139-144

Fitzgerald PH (1983) Premature centromere division of the X chromosome. In: Sandberg AA (ed) Cytogenetics of the mammalian X chromosome, Part A. Liss, New York, pp 171-184

Fryns JP, Berghe H Van den (1983) Y-X translocations in man. In: Sandberg AA (ed) Cytogenetics of the mammalian X chromosome, Part B. Liss, New York, pp 245-251

Glover TW (l983) The fragile X chromosome: factors influencing its expression in vitro. In: Sandberg AA (ed) Cytogenetics of the mammalian X chromosome, Part B. Liss, New York, pp 415-430

Hellkuhl B, Chapelle A de la, Grzeschik K-H (1982) Different patterns of X chromosome inactivity in lymphocytes and fibroblasts of a human balanced X;autosome translocation. Hum Genet 60: 126-129

Herva R, Kaluzewski B, Chapelle A de la (1979) Inherited interstitial del(Xp) with minimal clinical consequences: with a note on the location of genes controlling phenotypic features. Am J Med Genet 3: 43-58

Howard-Peebles PN (1983) Conditions affecting fragile X chromosome structure in vitro. In: Sandberg AA (ed) Cytogenetics of the mammalian X chromosome, Part B. Liss, New York, pp 431-443

Latt SA (1979) Patterns of late replication in human X chromosomes. In: Vallet HL, Porter IH (eds) Genetic mechanisms of sexual development. Academic, New York, pp 305-326

Maraschio P, Fraccaro M (1983) Phenotypic effects of X-chromosome deficiencies. In: Sandberg AA (ed) Cytogenetics of the Mammalian Chromosome, Part B. Liss, New York, pp 359-369

Mattei JF, Mattei MG, Aumeras C, et al (1981a) X-linked mental retardation with the fragile X. A study of 15 families. Hum Genet 59: 281-289

Mattei MG, Mattei JF, Vidal I, et al (1981b) Structural anomalies of the X chromosome and inactivation center. Hum Genet 56: 401-408

Narahara K, Yabuuchi H, Kimura S, et al (1978) A case of a reciprocal translocation between and the Y and no. 1 chromosomes. Jinrui Idengaku Zasshi [Jpn J Hum Genet] 23: 225-231

Otto PG, Therman E (1982) Spontaneous cell fusion and PCC formation in Bloom's syndrome. Chromosoma 85: 143-148

Sarto GE, Therman E, Patau K (1973) X inactivation in man: a woman with t(Xq−;12q+). Am J Hum Genet 25: 262–270

Schempp W, Meer B (1983) Cytologic evidence for three human X-chromosomal segments escaping inactivation. Hum Genet 63: 171–174

Spowart G (1979) Reassessment of presumed Y/22 and Y/15 translocations in man using a new technique. Cytogenet Cell Genet 23:90–94

Summitt RL, Tipton RE, Wilroy Jr RS, et al (1978) X-autosome translocations: a review. In: Birth defects: original article series, XIV, No. 6C. The National Foundation, New York, pp 219–247

Therman E (1983) Mechanisms through which abnormal X-chromosome constitutions affect the phenotype. In: Sandberg AA (ed) Cytogenetics of the mammalian X chromosome, Part B. Liss, New York, pp 159–173

Therman E, Denniston C, Sarto GE, et al (1980) X chromosome constitution and the human female phenotype. Hum Genet 54: 133–143

Therman E, Patau K (1974) Abnormal X chromosomes in man: origin, behavior and effects. Humangenetik 25: 1–16

Therman E, Sarto GE (1983) Inactivation center on the human X chromosome. In: Sandberg AA (ed) Cytogenetics of the mammalian X chromosome, Part A. Liss, New York, pp 315–325

Therman E, Sarto GE, DeMars RI (1981) The origin of telocentric chromosomes in man: a girl with tel(Xq). Hum Genet 57: 104–107

Therman E, Sarto GE, Disteche C, et al (1976) A possible active segment on the inactive human X chromosome. Chromosoma 59: 137–145

Therman E, Sarto GE, Palmer CG, et al (1979) Position of the human X inactivation center on Xq. Hum Genet 50: 59–64

Therman E, Sarto GE, Patau K (1974) Center for Barr body condensation on the proximal part of the human Xq: a hypothesis. Chromosoma 44: 361–366

Trunca C, Therman E, Rosenwaks Z (1984) The phenotypic effects of small, distal Xq deletions. Hum Genet 68: 87–89

Wahlström J (1983) Translocations between the X and Y chromosomes in the human: their dynamics and effects. In: Sandberg AA (ed) Cytogenetics of the mammalian X chromosome, Part B. Liss, New York, pp 253–260

Wolf U (1983) X-linked genes and gonadal differentiation. Differentiation 23, Suppl: 104–106

Wyss D, DeLozier CD, Daniell J, et al (1982) Structural anomalies of the X chromosome: personal observation and review of non-mosaic cases. Clin Genet 21: 145–159

Zakharov AF, Baranovskaya LJ (1983) X-X chromosome translocations and their karotype-phenotype correlations. In: Sandberg AA (ed) Cytogenetics of the mammalian X chromosome, Part B. Liss, New York, pp 261–279

Zuffardi O (1983) Cytogenetics of human X/autosome translocations. In: Sandberg AA (ed) Cytogenetics of the mammalian X chromosome, Part B. Liss, New York, pp 193–209

XXI
Numerically Abnormal Chromosome Constitutions in Humans

Abnormalities of Human Chromosome Number

As described in Chapter II, the chromosome number may change in two different ways. Either the number of *sets* of chromosomes increases, resulting in polyploidy (a decrease, leading to haploidy, does not occur in man), or the number of *individual*, normal chromosomes changes, giving rise to aneuploidy.

The only types of polyploidy found in man are triploidy and tetraploidy and the overwhelming majority of these cases end up as spontaneous abortions. The only known autosomal monosomy is the extremely rare 21 monosomy, and of the trisomies only three—those for chromosomes 13, 18, and 21—occur with any appreciable frequency in liveborn children (cf. Hassold and Jacobs, 1984). However, trisomies for most autosomes are found in spontaneous abortions (cf. Boué et al, 1976, 1985; Carr and Gedeon, 1977).

Polyploidy

Polyploidy is a common phenomenon in plant evolution, and occasional polyploids arise in most plant species. In plants, a tetraploid may be even "bigger and better" than the original diploid, whereas higher polyploids, such as hexaploids and octoploids, are usually less successful. Apart from fishes, lizards, and amphibians, polyploidy plays almost no role in animal evolution. However, individual polyploids occur occasionally in a few animal groups. Among the vertebrates the only major group in which natural polyploidy arises frequently are the amphibians. Especially in salamanders, polyploidy has been found in nature and created experimentally by treating the fertilized eggs with either cold or heat shocks (cf. Fankhauser, 1945). Not only triploid, tetraploid, and pentaploid, but even haploid larvae are viable in newts and axolotls. Although cell size increases with increasing chromosome number, the size and appearance of the larvae from haploid to tetraploid remain about the same. The pentaploids show some abnormalities. Apparently an adjustment of the cell number compensates for increased cell size (Fankhauser, 1945). Among the birds, triploid chickens and turkeys have been described (cf. Niebuhr, 1974).

The situation is very different in mammals. As a rule, polyploidy is lethal and results in prenatal death. Why are the effects of polyploidy so disastrous in this group, when in many other organisms polyploidy is not only harmless but may have beneficial effects? One possible factor is the discrepancy between the chromosome numbers of the maternal and fetal parts of the placenta. This possibility is supported by the observation that in triploid human fetuses the chorion is often abnormal, showing hydatidiform changes (cf. Niebuhr, 1974).

Human Triploids

A triploid zygote may come about through various processes (cf. Niebuhr, 1974). Either the egg cell or the sperm may have an unreduced chromosome number as a result of restitution in either the first or second meiotic division; or the second polar body may reunite with the egg nucleus; or two sperms may penetrate and fertilize the same egg cell.

Triploids form one of the largest groups of *heteroploid* (abnormal chromosome numbers) spontaneous abortions (Table XXI.1), representing about 17 percent of them (cf. Carr and Gedeon, 1977). Only about 1/ 10,000 triploid zygotes is born alive, and even of these most die within a day (cf. Niebuhr, 1974). Four triploid infants have survived 2, 5, 7, and 7 months (cf. Schröcksnadel et al, 1982). In such rare cases, one always

Table XXI.1. Relative Frequencies of Different Types of Chromosome Anomalies in 1863 Chromosomally Abnormal Spontaneous Abortions.[a]

Chromosome Anomaly	%
Trisomy	52
45,X	18
Triploidy	17
Tetraploidy	6
Other (mainly translocations)	7

[a]Modified from Carr and Gedeon, 1977.

suspects hidden mosaicism, and indeed most triploid infants have turned out to be mosaics with a diploid cell line.

Hydatidiform Moles

Hydatidiform moles represent abnormal growth of the trophoblast. They may develop, although rarely, after a normal pregnancy; usually, however, the zygote is abnormal. In Western countries 1/2,000 pregnancies leads to the development of a mole, whereas in the Orient, for instance Taiwan, this rate is 1/200. Hydatidiform moles fall morphologically, histologically, and genetically into two groups: partial and complete moles (Vassilakos et al, 1977; Boué et al, 1985).

In partial moles an abnormal, usually triploid, embryo is present. As a rule the triploidy has come about through an egg cell being fertilized by two sperms (Jacobs et al, 1982).

In complete moles no embryo is found. Usually they arise through an "empty" egg being fertilized by an X-bearing sperm and a subsequent doubling of the paternal chromosome set. (The origin of the empty egg cells is not known.) In other words, the chromosome complement of complete moles is 46,XX, the two chromosome sets being identical. This was first shown by an analysis of chromosome polymorphisms (Kajii and Ohama, 1977) and confirmed by the determination of enzyme polymorphisms and HLA specificities (Jacobs et al, 1980). Cytologically complete moles resemble malignant tumors more than normal placenta, showing giant nuclei, which come about through endomitosis and endoreduplication (Fig. XIII.1) (Sarto et al, 1984).

A minority of hydatidiform moles have other chromosome constitutions such as trisomic, 45,X, or 48,XXXX. Some 4 percent of them have a 46,XY chromosome complement, which is the result of two sperms fertilizing an empty egg cell (cf. Dodson, 1983).

Moles have attracted special interest because more than 50 percent of choriocarcinomas arise from them. Thus 2.1 percent of the partial and 10 percent of the complete moles become malignant. Fortunately chemotherapy cures some 80 percent of these highly malignant tumors (cf. Dodson, 1983).

Human Tetraploids

Tetraploid zygotes are more rare than triploids among spontaneous abortions. Altogether five infants with apparently nonmosaic tetraploidy have been reported so far (cf. Scarbrough et al, 1984). The few other liveborn children have been diploid/tetraploid mosaics. It is not surprising that tetraploid zygotes are rarer than triploids, since tetraploids are more severely affected and there are far fewer mechanisms that give rise to them. The most probable origin is chromosome duplication in a somatic cell at a very early stage of development. Among the other possible origins, it is highly unlikely that a rare unreduced sperm would by chance fertilize an equally rare diploid ovum. The fertilization of one egg cell by three sperms is also a possibility.

Autosomal Aneuploidy

In a number of plant species, such as corn, barley, spinach, snapdragon (*Antirrhinum*), and the classic object of trisomy studies—Jimson weed (*Datura*)—trisomics for all the chromosomes have been created (cf. Burnham, 1962). Each trisomic differs from any of the others and from normal diploid plants in many phenotypic features affecting the mode of growth, the leaves, the flowers, and the fruits. Each chromosome obviously contains genes that, in the trisomic state, modify many organ systems of a plant.

As already mentioned, human trisomics for only three autosomes (13, 18, and 21) are born alive with any appreciable frequency. Trisomy is

Table XXI.2. Number of Trisomics for Different Autosomes in a Total of 669 Trisomic Spontaneous Abortions (cf. Lauritsen, 1982).

Chromosome	Number of trisomic abortions	Percentage
1	0	0.0
2	33	4.9
3	4	0.6
4	17	2.5
5	1	0.2
6	3	0.5
7	27	4.0
8	26	3.9
9	18	2.7
10	13	2.0
11	2	0.3
12	7	1.0
13	31	4.6
14	31	4.6
15	52	7.7
16	216	32.3
17	4	0.6
18	34	5.1
19	1	0.2
20	18	2.7
21	63	9.4
22	68	10.2
Total	669	100.0

found for a few others, but rarely. However, trisomics for these three, as well as for most other autosomes, are observed in spontaneous abortions (Table XXI.2) (cf. Boué et al, 1976, 1985; Carr and Gedeon, 1977; Lauritsen, 1982). Mosaics for a normal and a trisomic cell line are found for more autosomes than full trisomy. Apparently a trisomic cell line that would be lethal by itself is sometimes able to exist in a mosaic individual who also has a normal cell line (Table XXI.4). In addition, an almost infinite variety of abnormal phenotypes is created by partial trisomies for certain chromosome segments distributed among all the autosomes. They are discussed in Chapters XXII and XXIII.

Anomalies Caused by Chromosomal Imbalance

In this book only a few chromosome syndromes are described as examples, since a vast literature dealing with the phenotypes of such conditions already exists. Concise and well-illustrated descriptions of chromosomal and other syndromes are found in Smith (1976). The book by de Grouchy and Turleau (1984) is completely oriented toward clinical cytogenetics. A list of papers dealing with chromosome aberrations has been published by Borgaonkar (1984). The book by Schinzel (1984) is a combination of the two previous books containing descriptions of chromosome syndromes and an extensive literature list. Syndrome descriptions are also given in Hamerton (1971), de Grouchy (1974), and Makino (1975). *New Chromosomal Syndromes* (Yunis, 1977) contains reviews on the phenotypes caused by imbalance for whole chromosomes and chromosome segments.

It has been repeatedly stressed that no phenotypic anomaly is exclusive to any chromosome syndrome, but rather that each of them is characterized by a combination of symptoms (cf. Smith, 1977). In the main this is true, although there are anomalies that are so rare that they can be regarded as characteristic of one or at most a couple of syndromes (cf. Lewandowski and Yunis, 1977). For example, the cat cry in infants with a 5p deletion is a specific trait. The cat cry syndrome is also characterized by prematurely gray hair. Polydactyly of hands or feet is found in patients with trisomy for chromosome 13, as is persistence of fetal hemoglobin (HbF) and abnormal projections in the neutrophils. Defects of the forebrain (holoprosencephaly) are also typical of 13-trisomic infants. The position of fingers, with the second overlapping the third, and the fifth the fourth, is a characteristic feature of the 18 trisomy syndrome, but is also found in some other syndromes.

In contrast to such specific symptoms, many anomalies are common to most chromosomal syndromes. Thus imbalance for even a small autosomal segment causes mental retardation. This is not surprising since the central nervous system is the most complicated of all organ systems, and therefore even a slight genetic imbalance has a deleterious effect on it. Furthermore, a high proportion of chromosomal syndromes is characterized by seizures of various kinds.

Growth retardation seems to be almost ubiquitous in chromosomal syndromes, the only exceptions being trisomy for chromosome 8 and for the short arm of chromosome 20 (cf. Francke, 1977). In many chromosomal syndromes, the failure to thrive, which is also reflected in abnormally low birth weight, is so extreme that they are essentially lethal conditions; the few infants who survive birth can be regarded as belated miscarriages (Smith, 1977). For instance, often 18-trisomic infants do not gain any weight after birth.

Heteroploidy seems to affect organ systems according to their complexity. Thus anomalies of the cardiovascular system are characteristic of a great variety of chromosomal syndromes. Ventricular septal defect is found in 47 percent of infants with 13 trisomy, in 34 percent of 18-trisomic patients, and in 20 percent of 21-trisomic children (Lewandowski and Yunis, 1977). A whole spectrum of other heart defects is also present in chromosomal syndromes.

One prerequisite for normal sex development is an XX or XY sex chromosome constitution. However, many autosomal syndromes exhibit defects in this system. For instance, infants with 13 trisomy often show cryptorchidism or a bicornuate uterus. The incidence of such effects is probably underestimated, since many infants with chromosomal syndromes die early, and their mature phenotypes remain unknown.

That no two persons have identical dermatoglyphics indicates that a great number of genes must be involved in their determination. Different constellations of dermatoglyphic features are characteristic of the various chromosomal syndromes. Therefore the genes responsible for dermatoglyphics are probably distributed on many chromosomes. As with other phenotypic anomalies, no single feature is specific for one chromosomal syndrome, but rather each condition is characterized by a combination of features. Individual features associated with specific syndromes also occur occasionally in the dermatoglyphics of normal persons.

The imbalance for even the smallest chromosome segment recognizable under the light microscope, excluding the heterochromatic regions, causes a variety of symptoms affecting many organ systems. This reflects the fact that even a small segment contains hundreds of genes. However, imbalances for *different* chromosome segments of about the *same length* may affect the phenotype to very different degrees. Obviously monosomy or trisomy is much more damaging for some genes than it is for others. It is also probable that the density of genes is higher in the Q-dark segments than in the bright ones, since trisomy or monosomy for them has much more drastic effects (cf. Kuhn et al, 1985).

The phenotypes of the more common chromosome syndromes are well defined by now. However, the mechanisms through which chromosomal imbalance exerts its influence are almost totally unknown. One possible pathway is that cells with different chromosome constitutions do not divide at the same rate (Paton et al, 1974). It is also known that in many mosaics the normal cell line has a selective advantage (cf. Nielsen, 1976). An abnormal division rate of cells may, in turn, change the growth rates of different tissues and organs, which might result in some of the phenotypic anomalies characteristic of chromosomal syndromes. The developmental study of chromosomally abnormal embryos is an interesting field of research that is only beginning (cf. Gropp et al, 1976).

13 Trisomy (D₁ Trisomy, Patau's Syndrome)

The chromosomes of an infant with 13 trisomy (Figs. XXI.1 and XXI.2) were first studied in our laboratory (Patau et al, 1960), and therefore this syndrome is described in detail as an example of trisomy syndromes. The extra D chromosome, which is the cause of the syndrome, was called D₁

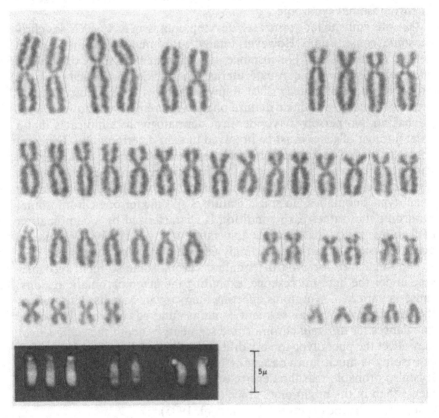

Figure XXI.1. Karyotype of a 13-trisomic male (orcein staining). Insert: Q-banded D-group with three chromosomes 13.

Figure XXI.2. First D-trisomic child described with anophthalmia, hare lip, six toes, retroflexed thumbs, and hemangiomata (Patau et al, 1960).

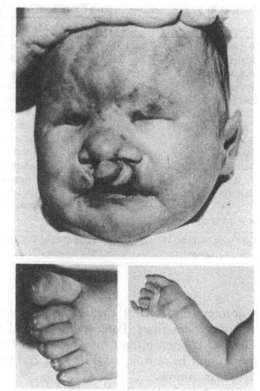

by Patau et al (1960) because it was thought that trisomics for the other two D chromosomes might be discovered later. Most 13-trisomic zygotes end up as spontaneous abortions (Table XXI.2), their frequency being 100 times greater in abortions than in liveborns.

The incidence of trisomy 13 (47, +13) in liveborns is, with 95 percent confidence, between 1/21,000 and 1/7,000, the most likely frequency being about 1/12,000 (cf. Hook, 1980). Increased maternal age is a factor in 13 trisomy, as it is for the other full trisomies. Even those 13-trisomics who survive birth have a limited life expectancy; about 45 percent die within the first month, 90 percent are dead before six months, and fewer than 5 percent reach the age of 3 years (cf. Gorlin, 1977; Niebuhr, 1977). However, two exceptional 13-trisomic patients have been reported to have lived much longer—11 and 19 years, respectively (Redheendran et al, 1981).

Infants with 13 trisomy are severely retarded mentally and are often deaf. Various degrees of forebrain defect (holoprosencephaly) are common (Table XXI.3). Eye anomalies range from anophthalmia to microphthalmia (Fig. XXI.2), often combined with coloboma of the iris (fissure of the iris). Cleft lip or cleft palate, or both are also characteristic of this

Table XXI.3. Frequency of Main Symptoms in 13 Trisomy.[a]

Symptom	% of cases	Symptom	% of cases
Severe retardation	100	Hemangiomata	73
Deafness	53	Polydactyly	78
Microcephaly	59	Heart disease	76
Hypertelorism	93	Renal anomaly	53
Epicanthic folds	52	Bicornuate uterus	43
Microphthalmia or anophthalmia	78	Arhinencephaly (absent olfactory bulbs)	71
Coloboma	35	Simian crease	64
Harelip	55	Distal triradius	77
Cleft palate	65	Elevated fetal hemoglobin (HbF)	>50
Malformed ears	81	Abnormal projections of neutrophils	>50
Low-set ears	11		

[a]Data from Niebuhr, 1977; other authors give somewhat different frequencies.

syndrome (Fig. XXI.2). Capillary hemangiomata and scalp defects have often been described. Polydactyly (Fig. XXI.2) was already mentioned. The heel is often prominent, and the feet exhibit a rocker bottom. The thumb may be retroflexed (Fig. XXI.2). Different types of heart anomalies are common.

Table XXI.3 gives the frequencies of the most common clinical anomalies in 13 trisomy (Niebuhr, 1977; see also de Grouchy, 1974; Hodes et al, 1978). The incidence of the various symptoms shows a wide range of variation from severe mental retardation, which is always present, to low-set ears, which have been reported in 11 percent of the cases. The considerable phenotypic variation displayed by 13-trisomic infants probably reflects the differences in the allele content of the three homologous chromosomes. The variation in the symptoms is also caused by the different ages of the patients, some symptoms vanishing and others becoming apparent with increasing age.

As in other mosaics, a normal cell line dilutes the effect of the trisomic cells, and as a result mosaic individuals have fewer and less severe anomalies than do full trisomics. In most normal trisomy mosaics, the abnormal cell line simply has an extra chromosome 13. However, an effectively trisomy mosaic child with an unusual karyotype was described by Therman et al (1963). In about 45 percent of the cells that had a 46,XX chromosome constitution, one D_1 chromosome was telocentric, whereas in the remaining cells this chromosome was replaced by an isochromosome (Chapter XVI). It was subsequently shown by autoradiography and by banding that this interpretation was correct. The 4-year-old girl was less retarded and had fewer anomalies than fully trisomic individuals.

Partial trisomy for different parts of chromosome 13 is also found (cf. de Grouchy, 1974; Niebuhr, 1977). As a rule these individuals have only some of the anomalies typical of 13 trisomy; partial trisomy for different segments shows specific combinations of symptoms. The observations of these specific symptoms are used for trisomy mapping of chromosome 13. According to de Grouchy (1974), 80 percent of 13 trisomy patients have the chromosome constitution $47,+13$, whereas the rest are either mosaics or have a Robertsonian translocation t(13qDq). The same chromosomes for which *trisomy* is viable are usually viable as *partial monosomics*. This is also true of chromosome 13, in which deletions are used for monosomy mapping (cf. Noel et al, 1976).

21 Trisomy Syndrome (Down's Syndrome, Mongolism)

This condition, which is the least severe of the autosomal trisomy syndromes, is described in detail in the books and reviews previously mentioned; in addition, entire volumes have been dedicated to various aspects of it (cf. Penrose and Smith, 1966; Lilienfeld, 1969). The chromosomal cause of Down's syndrome was discovered by Lejeune et al (1959). The syndrome is by far the most frequent of the autosomal trisomy syndromes, the estimates of its incidence ranging from 1/500 to 1/1,000 newborns, the usual estimate being 1/700. In roughly 95 percent of the cases the chromosome constitution is $47,+21$. In 2 percent of patients ascertained as probable 21 trisomics, mosaicism for a normal and a trisomic cell line is present. However, low-grade 21 trisomy mosaicism, which does not affect the phenotype, is sometimes found, especially in parents of 21-trisomic children. In about 3 percent of the patients the extra chromosome is attached to another chromosome, usually as a result of centric fusion to another acrocentric. Such Robertsonian translocations and their inheritance are discussed in Chapter XXIV.

In individuals with 21 trisomy, the probability of developing leukemia is increased 20-fold. Other causes of mortality are heart disease and infections, especially in the respiratory system.

Interestingly, a condition corresponding to 21 trisomy in both symptoms and chromosomal cause is found in the chimpanzee (McClure et al, 1969; Benirschke et al, 1974), the gorilla (Turleau et al, 1972), and the orangutan (Andrle et al, 1979).

18 Trisomy Syndrome (Edward's Syndrome)

Because of their severe failure to thrive, 18-trisomic infants often appear even more miserable than do 13 trisomics. The many anomalies characteristic of this syndrome have been described by de Grouchy (1974), Gorlin (1977), and Hodes et al (1978). The reported incidence of 18 trisomy varies between 1/3,500 and 1/7,000, or about 1/6,000 in the newborn. Of those born alive, 30 percent die within one month, and only 10 percent survive one year (cf. Gorlin, 1977). About 80 percent of the patients have straight trisomy, another 10 percent are mosaics, whereas the rest either are double trisomics for another chromosome or have a translocation (cf. de Grouchy, 1974).

Other Autosomal Aneuploidy Syndromes

Of the other trisomy syndromes (Table XXI.4), 22 trisomy has repeatedly been found in a nonmosaic condition. In their review of this syndrome. Hsu and Hirschhorn (1977) collected 18 cytologically confirmed cases from the literature. Especially interesting is the family studied by Uchida et al (1968), in which the mother was a mosaic and two children had full 22 trisomy. Interestingly, however, none of the carriers of a Robertsonian translocation involving chromosome 22 has produced a child trisomic for this chromosome (cf. Schinzel, 1984). The so-called *cat eye* syndrome is caused by an extra small chromosome, which seems to be a derivative of chromosome 22. In some cases this chromosome is apparently 22q−, in others an isodicentric of 22 with the breakpoint in the long arm (cf. Schinzel et al, 1981).

Table XXI.4. Human Chromosomes for which Trisomy and/or Trisomy Mosaicism Has Been Found.

Chromosome	Trisomy	Estimated Incidence of Trisomy in Newborns[a]	Trisomy Mosaicism
3	−	−	2 cases
7	+	not known	+
8	1 case	not known	+
9	+	not known	+
10	−	−	1 case
12	−	−	1 case
13	−	1/12,000	+
14	+	not known	+
15	1 case	not known	−
16	−	−	doubtful
18	+	1/6,000	+
19	−	−	2 cases
20	doubtful	−	doubtful
21	+	1/700	+
22	+	not known	+
X(XXX)	+	1/1,100 of same sex	+
X(XXY)	+	1/1,000 of same sex	+
Y(XYY)	+	1/1,000 of same sex	+

[a]In different studies the estimates vary to some extent.

Of the other autosomes, trisomy mosaicism for chromosome 3 has been found twice. One of the patients was discovered in our laboratory (unpublished). She was a 30-year-old, severely retarded woman with multiple anomalies. Possibly she had a higher percentage of trisomic cells when she was younger, but during development the normal cells have been selected for, constituting now 95 percent of the lymphocytes. The other case of trisomy 3 mosaicism was described in an infant (Metaxotou et al, 1981).

Full trisomy for chromosome 7 has been reported in a newborn who died after 2 days (Yunis et al, 1980). Chromosome 7 has also been

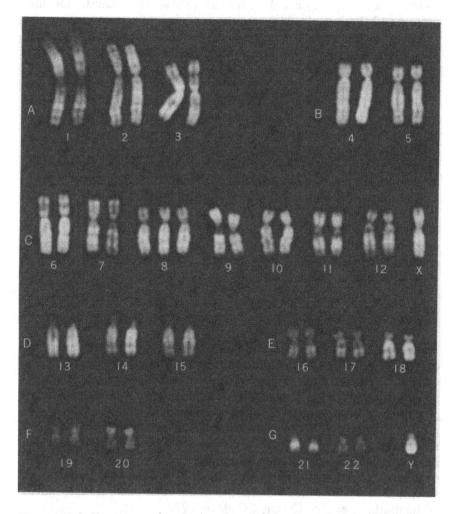

Figure XXI.3. Karyotype of an 8-trisomic cell from a mosaic male with 46,XY/ 47,XY,+8 (Q-banding).

involved in trisomy mosaicism, and an interesting family in which mother and daughter, both mentally ill, were such mosaics was found by De Bault and Halmi (1975). The most probable explanation for this family is that the zygote from which the daughter developed was trisomic, but later a normal cell line arose through loss of one chromosome 7. Trisomy for chromosome 8 is found repeatedly, but almost always in a mosaic state (Fig. XXI.3). Of the 61 patients reviewed by Riccardi (1977), only one was apparently nonmosaic.

Full trisomy for chromosome 9 has been reported in 20 cases and twice in a mosaic state (cf. Schinzel, 1984).

One case of trisomy mosaicism for chromosome 10 has been described (Nakagome et al, 1972) as well as one case of trisomy mosaicism for chromosome 12 (Patil et al, 1983).

Both mosaic and full 14 trisomy are virtually lethal conditions, and only a few such cases have been born alive (Johnson et al, 1979). One severely affected infant with full 15 trisomy has also been reported (Coldwell et al, 1981).

Although chromosome 16 trisomy is the most common trisomy in spontaneous abortions, its existence in liveborn infants is doubtful (cf. Schinzel, 1984).

Of the F group chromosomes, a couple of trisomy mosaics for chromosome 19 have been found; on the other hand, claims of chromosome 20 trisomy are not supported by sufficient evidence (cf. Schinzel, 1984).

Monosomy in liveborn infants has been established for only one autosome, chromosome 21. The *monosomy 21 syndrome* is extremely rare— only six cases having been reported (cf. Wisniewski et al, 1983)—and even in spontaneous abortions it is infrequent. Monosomy 21 has never arisen as a result of a Robertsonian translocation involving this chromosome.

Spontaneous Abortions

Table XXI.1 gives an estimate of the incidence of spontaneous abortions caused by different kinds of chromosome abnormalities. Carr and Gedeon (1977) estimate that 38 percent of spontaneous abortions are heteroploid. Obviously many factors affect such estimates, which indeed vary greatly in different studies (cf. Boué et al, 1985). A number of abortions must occur so early that they are unrecognizable, and even of the recognized ones, not all are amenable to chromosome analysis. Early abortions show more chromosome abnormalities than do later ones. The incidence of trisomic abortions increases with the maternal age. The frequency of polyploid as well as of 45,X abortions, on the other hand, is independent of the mother's age (cf. Carr and Gedeon, 1977).

Apart from the pooled trisomic abortions that have been found for all autosomes except chromosome 1, polyploidy (mainly triploidy) is the leading chromosomal cause of fetal loss. Of the individual classes, the lack of one sex chromosome (in other words, the chromosome constitution 45,X) is the most frequent. It is estimated that only 3 percent of the XO zygotes are born alive. Obviously the 45,X zygotes fall into separate classes of which one is lethal, whereas those who are born alive are not especially miserable.

Since monosomy is brought about by both chromosome loss and nondisjunction (whereas trisomy results only from the latter process), monosomics should be more frequent among spontaneous abortions than trisomics. In reality, monosomy is almost as rare in recognizable abortions as it is in liveborn children. Obviously monosomy for a chromosome—or for a chromosome segment—has much more deleterious effects than the corresponding trisomy. Apparently the monosomic zygotes, as a rule, die so early that they are not recognized as abortions.

In 669 spontaneous trisomic abortions reviewed by Lauritsen (1982) (Table XXI.2) not one was reported for chromosome 1, and even among the other chromosomes trisomic abortions were distributed very unevenly. As mentioned in Chapter V, Korenberg et al (1978) have put forward the hypothesis that chromosomes with especially gene-rich segments act as trisomy lethals (Patau, 1964) at a very early stage of gestation.

Another factor that may bias the observed numbers is the time at which the trisomic condition is lethal. For instance, the time of death caused by trisomy 13 ranges from early embryonic development to several years postnatally, whereas the lethal effect of chromosome 16 apparently occurs within a very limited period of gestation; the reported times vary from 22 to 31 days, at which stage chromosome studies are feasible (cf. Boué et al, 1976).

References

Andrle M, Fiedler W, Rett A, et al (1979) A case of trisomy 22 in *Pongo pygmaeus*. Cytogenet Cell Genet 24:1–6

Bault E De, Halmi KA (1975) Familial trisomy 7 mosaicism. J Med Genet 12: 200–203

Benirschke K, Bogart MH, McClure HM, et al (1974) Fluorescence of the trisomic chimpanzee chromosomes. J Med Prim 3:311–314

Borgaonkar DS (1984) Chromosomal variation in man: a catalog of chromosomal variants and anomalies, 4th edn. Liss, New York

Boué A, Boué J, Gropp A (1985) Cytogenetics of pregnancy wastage. In: Harris H, Hirschhorn K (eds) Advances in human genetics, Vol 14. Plenum, New York, pp 1–57

Boué J, Daketsé M-J, Deluchat C, et al (1976) Identification par les bandes Q et G des anomalies chromosomiques dans les avortements spontanes. Ann Génét 19:233–239

Burnham CR (1962) Discussions in cytogenetics. Burgess, Minneapolis

Carr DH, Gedeon M (1977) Population cytogenetics of human abortuses. In: Hook EB, Porter IH (eds) Population cytogenetics. Academic, New York, pp 1–9

Coldwell S, Fitzgerald B, Semmens JM, et al (1981) A case of trisomy of chromosome 15. J Med Genet 18:146–148

Dodson MG (1983) New concepts and questions in gestational trophoblastic disease. J Reprod Med 28:741–749

Fankhauser G (1945) The effects of changes in chromosome number on amphibian development. Q Rev Biol 20:20–78

Francke U (1977) Abnormalities of chromosomes 11 and 20. In: Yunis JJ (ed) New chromosomal syndromes. Academic, New York, pp 245–272

Gorlin RJ (1977) Classical chromosome disorders. In: Yunis JJ (ed) New chromosomal syndromes. Academic, New York, pp 59–117

Gropp A, Putz B, Zimmermann U (1976) Autosomal monosomy and trisomy causing developmental failure. In: Gropp A, Benirschke K (eds) Current topics in pathology 62. Springer, Heidelberg, pp 177–192

Grouchy J de (1974) Clinical cytogenetics: autosomal disorders. In: Busch H (ed) The cell nucleus. Academic, New York, pp 371–414

Grouchy J de, Turleau C (1984) Clinical atlas of human chromosomes, 2nd edn. Wiley, New York

Hamerton JL (1971) Human cytogenetics, Vol II. Academic, New York

Hassold TJ, Jacobs PA (1984) Trisomy in man. Annu Rev Genet 18:69–97

Hodes ME, Cole J, Palmer CG, et al (1978) Clinical experience with trisomies 18 and 13. J Med Genet 15:48–60

Hook EB (1980) Rates of 47,+13 and 46 translocation D/13 Patau syndrome in live births and comparison with rates in fetal deaths and at amniocentesis. Am J Hum Genet 32:849–858

Hsu LYF, Hirschhorn K (1977) The trisomy 22 syndrome and the cat eye syndrome. In: Yunis JJ (ed) New chromosomal syndromes. Academic, New York, pp 339–368

Jacobs PA, Szulman AE, Funkhouser J, et al (1982) Human triploidy: relationship between parental origin of the additional haploid complement and development of partial hydatidiform mole. Ann Hum Genet 46:223–231

Jacobs PA, Wilson CM, Sprenkle JA, et al (1980) Mechanism of origin of complete hydatidiform moles. Nature 286:714–716

Johnson VP, Aceto T Jr, Likness C (1979) Trisomy 14 mosaicism: case report and review. Am J Med Genet 3:331–339

Kajii T, Ohama K (1977) Androgenetic origin of hydatidiform mole. Nature 268:633–634

Korenberg JR, Therman E, Denniston C (1978) Hot spots and functional organization of human chromosomes. Hum Genet 43:13–22

Kuhn EM, Therman E, Denniston C (1985) Mitotic chiasmata, gene density, and oncogenes. Hum Genet 70:1–5

Lauritsen JG (1982) The cytogenetics of spontaneous abortion. Res Reprod 14:3–4

Lejeune J, Turpin R, Gautier M (1959) Le mongolisme, premier example d'aberration autosomique humaine. Ann Génét 1:41–49

Lewandowski RC, Yunis JJ (1977) Phenotypic mapping in man. In: Yunis JJ (ed) New chromosomal syndromes. Academic, New York, pp 369–394

Lilienfeld AM (1969) Epidemiology of mongolism. Johns Hopkins, Baltimore

Makino S (1975) Human chromosomes. Igaku Shoin, Tokyo

McClure HM, Belden KH, Pieper WA, et al (1969) Autosomal trisomy in chimpanzee: resemblance to Down's syndrome. Science 165:1010–1012

Metaxotou C, Tsenghi I, Bitzos M, et al (1981) Trisomy 3 mosaicism in a live-born infant. Clin Genet 19:37–40

Nakagome Y, Iinuma K, Matsui I (1972) Trisomy 10 with mosaicism. A clinical and cytogenetic entity. Jap J Hum Genet 18:216–219

Niebuhr E (1974) Triploidy in man: cytogenetical and clinical aspects. Humangenetik 21:103–125

Niebuhr E (1977) Partial trisomies and deletions of chromosome 13. In: Yunis JJ (ed) New chromosomal syndromes. Academic, New York, pp 273–299

Nielsen J (1976) Cell selection in vivo in normal/aneuploid chromosome abnormalities. Hum Genet 32:203–206

Noel B, Quack B, Rethore MO (1976) Partial deletions and trisomies of chromosome 13; mapping of bands associated with particular malformations. Clin Genet 9:593–602

Patau K (1964) Partial trisomy. In: Fishbein M (ed) Second international conference of congenital malformations. International Medical Congress, New York, pp 52–59

Patau K, Smith DW, Therman E, et al (1960) Multiple congenital anomaly caused by an extra autosome. Lancet i:790–793

Patil SR, Bosch EP, Hanson JW (1983) First report of mosaic trisomy 12 in a liveborn individual. Am J Med Genet 14:453–460

Paton GR, Silver MF, Allison AC (1974) Comparison of cell cycle time in normal and trisomic cells. Humangenetik 23:173–182

Penrose LS, Smith GF (1966) Down's anomaly. Little, Brown, Boston

Redheendran R, Neu RL, Bannerman RM (1981) Long survival in trisomy–13-syndrome: 21 cases including prolonged survival in two patients 11 and 19 years old. Am J Med Genet 8:167–172

Riccardi VM (1977) Trisomy 8: an international study of 70 patients. In: Birth defects: original article series, XIII, No. 3C. The National Foundation, New York, pp 171–184

Sarto GE, Stubblefield PA, Lurain J, et al (1984) Mechanisms of growth in hydatidiform moles. Am J Obstet Gynecol 148:1014–1023

Scarbrough PR, Hersh J, Kukolich MK, et al (1984) Tetraploidy: A report of three live-born infants. Am J Med Genet 19:29–37

Schinzel A (1984) Catalogue of unbalanced chromosome aberrations in man. De Gruyter, Berlin

Schinzel A, Schmid W, Fraccaro M, et al (1981) The "cat eye syndrome": dicentric small marker chromosome probably derived from a no. 22 (tetrasomy 22pter→q11) associated with a characteristic phenotype. Hum Genet 57:148–158

Schröcksnadel H, Guggenbichler P, Rhomberg K, et al (1982) Komplette Triploidie (69,XXX) mit einer Überlebensdauer von 7 Monaten. Wien Klin Wochenschr 94:309–315

Smith DW (1976) Recognizable patterns of human malformation. 2nd edn. Saunders, Philadelphia

Smith DW (1977) Clinical diagnosis and nature of chromosomal abnormalities. In: Yunis JJ (ed) New chromosomal syndromes. Academic, New York, pp 55–58

Therman E, Patau K, DeMars RI, et al (1963) Iso/telo-D₁ mosaicism in a child with an incomplete D₁ trisomy syndrome. Portugal Acta Biol 7:211–224

Turleau C, Grouchy J de, Klein M (1972) Phylogénie chromosomique de l'homme et des primates hominiens (Pan troglodytes, Gorilla gorilla, et Pongo pygmaeus): essai de reconstitution du caryotype de l'ancetre commun. Ann Génét 15:225–240

Uchida IA, Ray M, McRae KN, et al (1968) Familial occurrence of trisomy 22. Am J Hum Genet 20:107–118

Vassilakos P, Riotton G, Kajii T (1977) Hydatidiform mole: two entities. Am J Obstet Gynecol 127:167–170

Wisniewski K, Dambska M, Jenkins EC, et al (1983) Monosomy 21 syndrome: further delineation including clinical, neuropathological, cytogenetic and biochemical studies. Clin Genet 23:102–110

Yunis E, Ramirez E, Uribe JG (1980) Full trisomy 7 and Potter syndrome. Hum Genet 54:13–18

Yunis JJ (ed) (1977) New chromosomal syndromes. Academic, New York

XXII
Structurally Abnormal Human Autosomes

Structurally Abnormal Chromosomes

In contrast to full trisomy and monosomy, which in liveborn infants are limited to a few autosomes, the variety of structurally abnormal chromosomes is almost infinite. This is to be expected, since chromosomes may break at almost any point, and the broken ends may join randomly to form new combinations. Chromosome breaks occur in meiocytes as well as in somatic cells. The only limitation to the variety of structurally abnormal chromosomes is their possible lethal effect on the individuals or cells carrying them. Deletions, duplications, ring chromosomes, inversions, and translocations cause partial trisomy and/or monosomy syndromes which have been described for all chromosome arms (cf. de Grouchy and Turleau, 1984; Schinzel, 1984). It would be impossible to review the entire field of structural chromosome abnormalities even in a much more comprehensive book than this one; therefore, only the main classes of abnormal chromosomes are described herein, illustrated by a few examples of the resulting syndromes. Reciprocal and Robertsonian translocations and their segregation are discussed in Chapters XXIII and XXIV.

Chromosomal Polymorphisms

Chromosomal polymorphisms or *heteromorphisms* are structural variants of chromosomes that are widespread in human populations and have no effect on the phenotype, even in their most extreme forms. The apparent harmlessness of chromosomal polymorphisms led to the conclusion that segments displaying such variants must be heterochromatic. With band-

ing techniques we now know that these polymorphisms represent constitutive heterochromatin, often at the centric regions of the chromosomes.
In a comparison of the incidence of such chromosome variants in 200
mentally retarded patients and the same number of normal controls, no
difference was found between the two groups (Tharapel and Summitt,
1978). However, different results have been obtained in other studies.

   The polymorphic segments vary in size, position (through inversion),
and staining properties. This has been shown with C-banding, Q-banding,
R-banding with acridine orange, and various combinations of fluorescent
dyes such as DAPI and DIPI (Verma and Dosik, 1980; Schnedl et al,
1981). The following chromosome segments display polymorphism: (1)
Q-bright distal end of the Y chromosome, which varies from zero to three
or four times its average length (Fig. XVII.1b), extreme variants being
very rare. (2) The short arms and satellites of acrocentric chromosomes,
which vary both in size and in fluorescent characteristics. The length of
the satellite stalks where the ribosomal RNA genes are situated vary also
in length, or they may be absent on certain chromosomes (cf. Mikelsaar
et al, 1977). (3) A Q-bright centric region may or may not be present in
chromosomes 3, 4, 13, and 22. (4) Centric heterochromatin as revealed
by C-banding shows a wide range of variation, especially in chromosomes

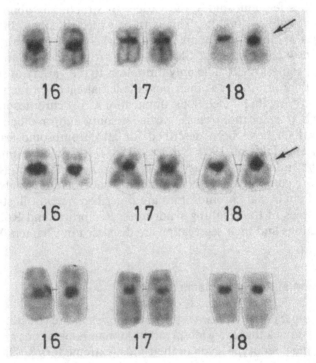

Figure XXII.1. E group from father, daughter, and mother. The exceptionally large
C-band in chromosome 18 in father and daughter was used for determination of
paternity (courtesy of HD Hager and TM Schroeder).

1 (Fig. V.1f), 9, and 16, but some variation is seen in many other chromosomes, such as 12, 17, and 21 (Mayer et al, 1978) and 19 (Trunca Doyle, 1976). It is likely that all the C-bands vary to some extent (cf. Craig-Holmes, 1977) (Fig. XXII.1).

Possibly the variation of the distal segment of the Y and of the C- bands in chromosomes 1, 9, and 16 is not continuous, but the heterochromatin consists of "blocks" whose number may vary (Madan and Bobrow, 1974; Magenis et al, 1977).

In addition to changes in size, heterochromatic regions are prone to other structural aberrations. Numerous inversions that involve the centric heterochromatin of chromosomes 1 and 9 (Fig. XXIV.1) and the Q-bright centric band on chromosome 3 have been described (cf. Kaiser, 1980).

Most of the fragile regions also represent chromosome polymorphisms.

Quantitative Comparisons

Quantitative comparisons of the sizes of C-bands are difficult, even within one study (cf. Verma and Dosik, 1980). This is also true of determining inversions in these regions. It is therefore not surprising that the estimates of the incidence of heterochromatic size variants in chromosome 9 range from 0.1 percent to 12.5 percent (cf. Sanchez and Yunis, 1977), and the frequency of inversions involving the same region ranges from 0.7 percent to 11.3 percent (cf. Sanchez and Yunis, 1977). Although populations may indeed vary with respect to the incidence of heterochromatic variants, clearly the strikingly different results are caused mainly by the nonuniform criteria used, combined with technical difficulties. The final results on the incidence of heterochromatic variants are not yet in.

Chromosomal polymorphisms are used as markers in gene mapping and cell hybridization studies. They can be helpful in determining zygosity in twins and in settling disputes about paternity (Fig. XXII.1). They also show which parent provided an extra chromosome in trisomics or an extra chromosome set in triploids. In addition, polymorphisms have played an important role in cytogenetic studies of hydatidiform moles. Finally the origin of the different cell lines in mosaics and chimeras, including persons with a bone marrow transplant, can be determined with the help of polymorphisms.

Pericentric Inversions

Inversions fall into two groups: in pericentric inversions the breaks are in opposite arms, in paracentric ones in the same arm. Pericentric inversions are often discovered because they change the position of the centromere.

Kaiser (1980) has listed 146 major pericentric inversions in humans. They have been found in all chromosomes except 12, 17, and 20. The different chromosomes are involved nonrandomly; for instance, inv(9) makes up some 40 percent of them. The breakpoints are also nonrandom. For example, 7 out of 10 inversions involving chromosome 2 showed breakpoints p11 and q13 (Leonard et al, 1975; Kaiser, 1980). Interestingly, 2q13 is also a fragile site.

The estimated incidence of inversions in human populations varies greatly in different studies: a number often mentioned is 1 percent. This may depend in part on the difficulty of drawing a line between polymorphisms and "real" inversions that may result in the birth of abnormal offspring.

In meiosis, short heterozygous inversions remain unpaired. A large inverted segment usually forms a loop to pair with its homologue, or it may pair straight, leaving the now nonhomologous ends unpaired. Crossing-over in a pericentric inversion leads to a deletion and duplication (Fig. XXII.2), a phenomenon called "aneusomie de recombinaison" by the French authors. Crossing-over in a paracentric inversion may produce a dicentric and an acentric chromosome. The complicated relationships of various types of crossing-over in inversions and the chromosome configurations arising from them are reviewed in many cytogenetics textbooks, for instance, in Burnham (1962).

It is surprising how few of the inversions in man seem to lead to reproductive trouble—either to the birth of recombinant offspring or to partial sterility (cf. Moorhead, 1976). For small inversions this is understanda-

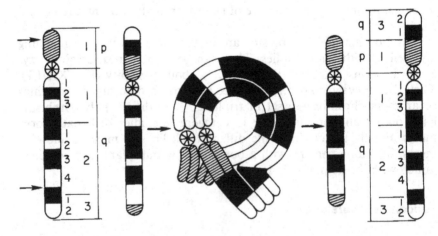

Figure XXII.2. Breakpoints in a pericentric inversion of chromosome 14 and two types of abnormal chromosomes (one with duplication, the other with deletion) resulting from crossing-over in an inversion (Trunca and Opitz, 1977).

ble, since they rarely undergo legitimate pairing. However, even larger inversions are often genetically benign.

Trunca and Opitz (1977) in a study of a woman with inv(14) and her abnormal child, who had a duplication in the same chromosome (Fig. XXII.2), review the factors promoting the incidence of abnormal offspring. The risk for an inversion carrier is determined by the probability of either type of recombinant offspring surviving birth. Trunca and Opitz (1977) divided the pericentric inversions into those involving less than one-third of the chromosome and those involving more. All families in the latter group had at least one abnormal child, those in the former none. The obvious explanation is that the longer the inversion relative to the chromosome, the greater the probability that crossing-over will occur. At the same time, the duplicated and deleted segments become smaller, as less of the chromosome remains outside the inverted segment. Both factors promote the birth of abnormal but viable offspring.

Of recombinant offspring, the type with the larger duplication and smaller deletion is usually more viable. However, in some unfortunate families both types of abnormal children have been born alive (for example, Vianna-Morgante et al, 1976).

Paracentric Inversions

Far fewer paracentric than pericentric inversions have been found. Djalali et al (1984) list 38 such inversions, to which one more should be added (Valcárcel et al, 1983). Paracentric inversions have been found in chromosomes 1, 3, 5, 6, 7, 8, 12, 13, 14, and the X. This nonrandomness is reflected also in the chromosome arms involved—3p, 6p, and 7q each showing 5 paracentric inversions. Of the inversions in 7p, 4 had breakpoints at p13 and p25.

The relative rarity of the observed paracentric inversions is probably caused by several factors. First, it is less probable for two breaks to be in the same arm than on different arms. Paracentric inversions are also more difficult to detect, since they do not change the arm ratio. Furthermore, crossing-over in a paracentric inversion leads to a dicentric and an acentric chromosome; this results in few viable offspring, through whom chromosomal abnormalities are usually ascertained.

Deletions or Partial Monosomies

Chromosome deletions or partial monosomies may be divided into two groups, pure deletions and deletions combined with a duplication, which usually result from reciprocal translocations. Deletions affecting all chromosome ends as well as many interstitial segments have been found in liveborn children (cf. de Grouchy and Turleau, 1984; Schinzel, 1984). Certain deletions are relatively frequent, for instance 4p−, 5p−, 9p−, 11p−, 11q−, 13q−, 18p−, and 18q−, whereas others have been found only a few times.

As with sex chromosome anomalies, monosomy for an entire autosome or segment thereof has much more serious phenotypic consequences than the corresponding trisomy. Monosomy syndromes also display a wider range of phenotypic variation than those caused by trisomy. A good example of such variation is provided by the family described by Uchida et al (1965). A retarded, apparently nonmosaic woman with a deletion of the whole 18p had two children with the same chromosome anomaly. The viability of the mother and the first child did not seem to be impaired, although they displayed a number of anomalies. The second child, however, who was much more severely affected, died on the third day after birth.

With prophase banding it is now possible to detect very small changes in chromosomes. Small deletions have been found to be the cause of several defined syndromes, such as Wilms' tumor and aniridia, retinoblastoma, and probably Prader Willi-syndrome (Chapter XXVI).

It is safe to predict that when conditions previously thought to be caused by a mutant gene are studied with high resolution techniques, a proportion of them will be found to depend on small deletions (duplications of equal size probably do not affect the phenotype noticeably). Especially hopeful in this respect would be presumed genes whose inheritance shows some oddity, for example a recessive syndrome that also affects a cousin.

Cri du Chat (Cat Cry) Syndrome

By far the most common of the deletion syndromes, and the one most extensively studied, is the so-called *cri du chat syndrome*, caused by a partial deletion of 5p. The incidence of this condition in infants is estimated as 1/45,000, and its frequency among the mentally retarded as 1.5/1000 (cf. Niebuhr, 1978b).

Niebuhr (1978b) reviewed 331 patients with cri du chat syndrome. One of the most consistent symptoms is the catlike infant cry which, however, is modified with age. The cry seems to be caused mainly by defects of the central nervous system (Schroeder et al, 1967). Niebuhr (1978b) lists 50 symptoms characteristic of this syndrome, many of which are probably interdependent. In addition to the characteristic cry, the most common abnormalities according to Smith (1976) are: severe mental retardation (100 percent), the IQ in adults being less than 20; hypotonia (100 percent), which in adults turns into hypertonia; microcephaly (100 percent); round face in children (68 percent); hypertelorism (94 percent); epicanthic folds (85 percent); downward slanting of palpebral fissures (81 percent); strabismus (61 percent); low-set or poorly developed ears (58 percent); heart disease of various types (30 percent); and characteristic dermatoglyphics (about 80 percent).

In Fig. XXII.3 three cri du chat patients representing different age groups are portrayed. The differences in the severity of corresponding monosomy and trisomy syndromes are well demonstrated by cases in which segregation in translocation families has produced both cri du chat and its "countertype" offspring, who have partial trisomy for 5p. As a rule, the latter have fewer symptoms, suffering mainly from mental retardation (cf. Yunis et al, 1978).

In a family described by Opitz and Patau (1975), two carriers had a t(5p−;12q+). Six infants with numerous anomalies and partial trisomy for 5p were born in two generations. The severity of the syndrome and the lack of any cri du chat offspring probably depend on the lethality when most of 5p is deleted, as shown by the diplochromosome karyotype of a carrier (Fig. XXII.4).

The size of the deletion of 5p causing the cri du chat syndrome varies from very small to about 60 percent of the length of the chromosome arm. It was observed years ago that the length of the deletion showed little correlation with the severity of the syndrome. This became understandable when Niebuhr (1978a), in a detailed cytological study of 35 patients, using both Q-banding and R-banding, showed that the critical segment (whose absence caused most of the symptoms) was a very small region around the middle of 5p15 (Fig. XXII.5). Twenty-seven of the 35 patients apparently had a terminal deletion, whereas four were the result of translocations, two of which were familial. Similar results have been obtained

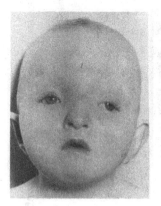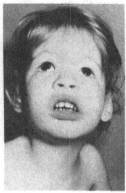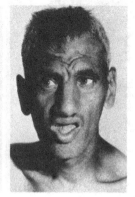

Figure XXII.3. Three cri du chat patients (infant, 4 years, 40 years), showing moon face, hypertelorism, antimongoloid slant of the eyes, downward slant of the mouth, and low-set ears (courtesy of R Laxova).

Figure XXII.4. Diplochromosome karyotype of a balanced carrier of t(5p−;12q+) (confirmed with Q-banding).

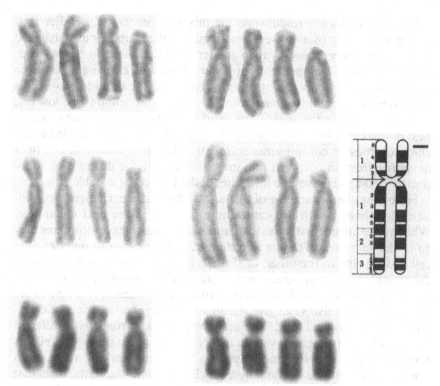

Figure XXII.5. The B chromosomes from two cells each of three cri du chat patients; length of the deletion is different in each patient. The lack of middle part of 5p15 (marked in diagram) is supposed to cause all the symptoms.

in larger populations (cf. Niebuhr, 1978b): in 80 percent of the patients the syndrome was caused by a deletion, in 10 percent one of the parents had a translocation, and another 10 percent of the patients showed other chromosome abnormalities, such as rings, de novo translocations, and mosaicism. All in all, 12 percent of the cases were familial.

Ring Chromosomes

Ring chromosomes have been found for all human chromosomes. Since a ring involves a deletion at each end of the chromosome, the resulting phenotypes overlap with deletion syndromes for the same chromosome. However, it has been observed repeatedly that the phenotypes of carriers of ring chromosomes vary greatly from mentally retarded with multiple anomalies to apparently normal, fertile persons (Hecht and Vlietinck, 1973). Sometimes even persons with apparently identical ring chromosomes have nothing in common phenotypically (for example, Zackai and Breg, 1973; Dallapiccola et al, 1977).

Several factors are responsible for this phenomenon. In earlier textbooks, ring chromosomes were described as either stable or unstable, and indeed some rings show a much greater stability than others. However, since the ubiquitous presence of sister chromatid exchanges has been established, one wonders why any ring chromosome should show even a slight degree of stability. One reason for this phenomenon is the low rate of naturally occurring sister chromatid exchanges and the nonrandomness of their location. Consequently they may be extremely rare in certain chromosomes or chromosome segments. Also the viability of cell lines deviating from the original one may vary greatly for different chromosomes. Perhaps factors like these create the illusion of a stable ring.

Following the occurrence of one sister chromatid exchange, a continuous double ring with one centromere is found in the next metaphase. When the centromere divides in anaphase, the daughter centromeres may go to the same pole, leading to a double, dicentric ring in one daughter cell and no ring in the other; or if the centromeres go to opposite poles, the daughter cells may obtain unequal rings. If the chromosome is twisted when the chromatids join, a new dicentric will be formed. It is easy to see how such mechanisms can give rise to an almost infinite variety of derivatives of the original ring. The main variants are: rings with different numbers of centromeres (even an octocentric ring in a polyploid cell has been described by Niss and Passarge, 1975), interlocking rings, rings consisting of variable combinations of chromosome segments, rings that have opened up, interphase-like chromosomes in metaphase, and several rings in one cell (cf. Hoo et al, 1974).

It is interesting that in many ring chromosome carriers, cells or cell lines in which the ring is missing have been observed. However, what is unexpected is that cell lines that are monosomic for certain chromosomes, for instance 6 (van den Berghe et al, 1974) or 10 (Lansky et al, 1977), are viable, although they have never been observed in persons without a ring chromosome. One cannot help wondering whether the cells with double rings might somehow compensate for the monosomic cells.

Causes of the great phenotypic variation in carriers of rings involving the same chromosome include: (1) the size of the original deletions, (2) the rate of sister chromatid exchanges in the ring, and (3) the viability of the cell lines with aberrant ring chromosomes. Figure XXII.6 (see also Fig. VII.2g) illustrates an r(9) in which the deletions obviously were minute. However, the severe retardation and multiple congenital anomalies in this patient (originally she had been diagnosed as having the Cornelia de Lange syndrome) are easy to understand (ML Motl, unpublished), as roughly 25 percent of her cells had a double or otherwise abnormal ring chromosome. Her phenotype falls into the—admittedly variable—range of conditions found in carriers of an r(9) (cf. Nakajima et al, 1976; Inouye et al, 1979).

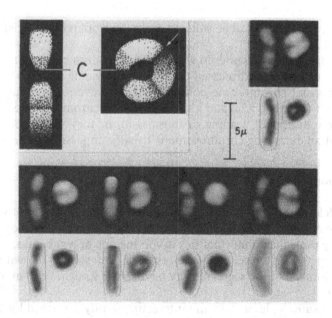

Figure XXII.6. Ring chromosome 9 with very small deletions and the normal 9 and r(9) from 5 cells (courtesy of ML Motl).

Although most persons with a ring chromosome are both mentally and somatically severely affected, a number of apparently normal and even fertile carriers have been found. This obviously depends on the minute size of the deletions and the relative stability of the ring. Thus a 5-year-old girl with an r(4) was not even mentally retarded. However, she displayed short stature, a small head, and retarded bone age (Surana et al, 1971). More puzzling is the case of a nonretarded 9-year-old girl, described by Lansky et al (1977), who had very few other anomalies either. One-half of her lymphocytes showed an r(10), whereas the other half were monosomic for chromosome 10. Since only lymphocytes were studied, she may have been a mosaic with a normal cell line, which is also the most probable explanation for other cases that deviate from the expectation.

At least five apparently normal women with 46,XX,r(21) have been described (Matsubara et al, 1982; Schmid et al, 1983; Kleczkowska and Fryns, 1984; Rhomberg, 1984). One of them had a normal son with the same ring, and two others had Down's syndrome children with an extra r(21), one of them a mosaic. A further woman with 46,XX,r(14) and low normal intelligence had two retarded children and a therapeutically aborted fetus with the same ring (Riley et al, 1981). Finally a retarded and psychotic woman with an r(18) gave birth to a daughter, bearing the same chromosome, who became even more retarded and psychotic. In addition, the child's father suffered from schizophrenia (Christensen et al, 1970).

Apart from the two children with Down's syndrome mentioned above, in all the cases herein discussed the ring chromosome replaced a member of the normal chromosome complement. However, numerous patients have been described who had a small extra ring chromosome. Such small rings may disappear, leading to mosaicism, or they may open up. The origin of an extra ring chromosome is usually impossible to determine.

Insertions

While deletions require only one or two breaks and inversions two, *insertions* are necessarily the result of three breaks. Consequently they are considerably rarer than abnormalities involving one or two breaks. An insertion may occur either between two chromosomes or within one chromosome, and the segment may be inserted straight or in an inverted position.

As an example, let us take the interesting family described by Therkelsen et al (1973). In the carrier father a segment from 2p had been inserted into 2q. Two abnormal infants were born, representing different types of recombination products.

In another family in which a segment from chromosome 13 had been inserted in an inverted position into 3q, segregation could be followed in three generations (Toomey et al, 1978). Both the monosomic and the trisomic types of recombinant individuals, as well as carriers and normals, were observed.

Duplication or Pure Partial Trisomy

Partial trisomy that is not combined with a deletion either for the same chromosome or for a nonhomologous one is relatively rare. In a couple of cases, an abnormal individual has been found to have a de novo duplication (Vogel et al, 1978). Either of two mechanisms would give rise to such a tandem duplication: an insertion from the homologous chromosome or unequal sister chromatid exchange or crossing-over in meiosis or mitosis.

As a rule, crossing-over in an inversion leads to abnormal chromosomes that have both a duplication and a deletion. Only when the inversion involves an acrocentric chromosome and one break has taken place in the short arm will the recombinant chromosome have a pure duplication (Fig. XXII.2) (Trunca and Opitz, 1977).

One of the clearest examples of pure partial trisomy is provided by isochromosomes. However, such chromosomes are very rarely members of a 46-chromosome complement, since this would involve monosomy for the other chromosome arm. In consequence, apart from Xq and Yq, isochromosomes have been found only for 9p and 18q (Rodiere et al, 1977) and the acrocentrics, in which the lack of the short arm is not deleterious.

Isochromosomes or isodicentrics in addition to a normal chromosome complement have been described for very short arms, such as 18p (Nielsen et al, 1978) and 17p (Mascarello et al, 1983). In such cases the individual is naturally tetrasomic for the arm concerned. A chromosomally interesting child was described by Leschot and Lim (1979). She had a t(2q;5q), whereas 5p formed an isochromosome, which resulted in pure trisomy for this arm. A similar combination of translocation and iso-chromosome formation for 4p has also been found (André et al, 1976).

Trisomy for 9p is caused by either an isochromosome for this arm or an extra free 9p chromosome (cf. Biederman, 1979). Even tetrasomy for major parts of chromosome 9 seems to be compatible with at least a limited life span. The largest partial tetrasomy has been described in a patient with an extra isodicentric consisting of two chromosomes 9 attached, long arm to long arm (breakpoint in both q22), with the second centromere inactivated (Wisniewski et al, 1978). The infant was highly abnormal and lived for only a couple of hours. Four other cases of partial tetrasomy for the same chromosome are reviewed by Wisniewski et al (1978).

Small chromosomes that are extra to the normal chromosome complement are found both in normal and abnormal individuals. Mosaicism is common in them, and with increasing age the proportion of the cells with the extra chromosome often decreases. Although the exact origin of the small extra chromosomes in most cases remains uncertain, they can be divided into different categories. (1) Metacentric chromosomes with satellites at both ends, which do not affect the phenotype, obviously consist of two heterochromatic short arms of the acrocentrics. (2) The extra chromosome represents a deleted acrocentric chromosome with satellites on its short arms, or sometimes on both arms. Some such chromosomes are isodicentrics. (3) The extra chromosome consists of the centric region of a nonacrocentric chromosome. Obviously two factors determine the size of such chromosomes. On the one hand, very small chromosomes are usually lost; on the other, the longer the chromosome is, the more likely it is to have lethal effects. (4) An extra chromosome may be the result of a 1:3 segregation in a translocation carrier. (5) Isochromosomes have already been discussed. (6) As previously mentioned, small rings occur in addition to the normal chromosome complement. Sometimes they open up and resemble ordinary chromosomes.

Dicentric Chromosomes

Dicentric chromosomes, although rare, may constitute permanent members of human chromosome complements. Dicentrics may consist of two nonhomologous chromosomes or of two homologues. In the latter case they are usually broken at identical points; the resulting chromosome is an isodicentric.

Two factors permit the continued existence of dicentrics: If the two centromeres are close, as in Robertsonian translocations, the chromosome functions as a monocentric. However, the second and even more important phenomenon is the ability of human dicentric chromosomes to inactivate one centromere. This is expressed visually in the lack of a second primary constriction, although C-bands are present at both centromere sites. Although dicentric chromosomes are a negligible source of abnormal human phenotypes, centromere inactivation is a highly interesting phenomenon and may have played an important role in the evolution of mammalian karyotypes. This happened, for instance, when two chromosomes of the great apes fused to form human chromosome 2.

Dicentric chromosomes arise through a reciprocal translocation in G_1, or usually through segregation of an adjacent quadriradial (Chapter VII).

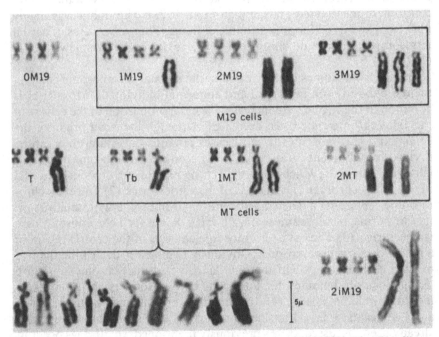

Figure XXII.7. Partial karyotypes of cells of a woman with 46,XX,t(6p;19p), illustrating the behavior of the translocation chromosome. Nomenclature: 0M19, "normal" cell; 1M19, cell with one large marker fragment; 2M19, cell with two markers; 3M19, cell with three markers; T 6;19 translocation chromosome; Tb translocation chromosome showing a chromatid break at the fusion point; 1MT cell carrying the translocation chromosome plus a marker fragment; 2MT translocation plus two markers. Chromosomes in brackets are examples of Tb from 10 cells; 2iM19 cell with two isofragments, each consisting of two chromosomes 6 (Drets and Therman, 1983).

Most isodicentrics consist of two X chromosomes (Chapter XX), but also Y isodicentrics are frequent. Dicentrics between nonhomologous chromosomes often involve the short arm of an acrocentric, which minimizes the deletion effect (Lambert et al, 1979).

Sometimes the chromosomes have joined at the telomeres (for example, Sarto and Therman, 1980). This is apparently true also of a t(6p;19p) in which the centromere of chromosome 6 is inactivated (Drets and Therman, 1983). This chromosome behaves exceptionally in that it has a tendency to break at the fusion point of the two chromosomes (1/733 cells). Since the centromere of the broken-off chromosome 6 is not reactivated, this chromosome behaves like an acentric fragment. These processes have led to complicated mosaicism. Figure XXII.7 illustrates partial karyotypes of 12 different cell types found among the 8,800 lymphocyte metaphases analyzed (other abnormalities are also present). As in the ring chromosome cases, surprisingly unbalanced cells seem to be able to divide, as demonstrated by the cell with two isofragments of chromosome 6, which thus in effect has 5 chromosomes 6.

References

André M-J, Aurias A, Berranger P de, et al (1976) Trisomie 4p de novo par isochromosome 4p. Ann Génét 19:127–131

Berghe H van den, Fryns J-P, Cassiman J-J, et al (1974) Chromosome 6 en anneau. Caryotype 46,XY,r(6)/45,XY,–6. Ann Génét 17:29–35

Biederman B (1979) Trisomy 9p with an isochromosome of 9p. Hum Genet 46:125–126

Burnham CR (1962) Discussions in cytogenetics. Burgess, Minneapolis

Christensen KR, Friedrich U, Jacobsen P, et al (1970) Ring chromosome 18 in mother and daughter. J Ment Defic Res 14:49–67

Craig-Holmes AP (1977) C-band polymorphism in human populations. In: Hook EB, Porter IH (eds) Population cytogenetics. Academic, New York, pp 161–177

Dallapiccola B, Brinchi V, Curatolo P (1977) Variability of r(22) chromosome phenotypical expression. Acta Genet Med Gemellol 26:287–290

Djalali M, Steinbach P, Barbi G (1984) Familial paracentric inversion inv(3)(q21q25.1). Ann Génét 27:41–44

Drets ME, Therman E (1983) Human telomeric 6;19 translocation chromosome with a tendency to break at the fusion point. Chromosoma 88:139–144

Grouchy J de, Turleau C (1984) Clinical atlas of human chromosomes, 2nd edn. Wiley, New York

Hecht F, Vlietinck RF (1973) Autosomal rings and variable phenotypes. Humangenetik 18:99–100

Hoo JJ, Obermann U, Cramer H (1974) The behavior of ring chromosome 13. Humangenetik 24:161–171

Inouye T, Matsuda H, Shimura K (1979) A ring chromosome 9 in an infant with malformations. Hum Genet 50:231–235

Kleczkowska A, Fryns JP (1984) Ring chromosome 21 in a normal female. Ann Génét 27:126–127

Kaiser P (1980) Pericentrische Inversionen menschlicher Chromosomen. Thieme, Stuttgart

Lambert JC, Ferrari M, Bergondi C, et al (1979) 18q− syndrome resulting from a tdic(14p;18q). Hum Genet 48:61–66

Lansky S, Daniel W, Fleiszar K (1977) Physical retardation associated with ring chromosome mosaicism: 46,XX,r(10)/45,XX,10−. J Med Genet 14:61–63

Leonard C, Hazael-Massieux P, Bocquet L (1975) Inversion pericentrique inv(2)(p11q13) dans des familles non apparentees. Humangenetik 28:121–128

Leschot NJ, Lim KS (1979) "Complete" trisomy 5p: de novo translocation t(2;5)(q36;p11) with isochromosome 5p. Hum Genet 46:271–278

Madan K, Bobrow M (1974) Structural variation in chromosome No. 9. Ann Génét 17-81–86

Magenis RE, Palmer CG, Wang L, et al (1977) Heritability of chromosome banding variants. In: Hook EB, Porter IH (eds) Population cytogenetics. Academic, New York, pp 179–188

Mascarello JT, Jones, MC, Hoyme HE, et al (1983) Duplication (17p) in a child with an isodicentric (17p) chromosome. Am J Med Genet 14:67–72

Matsubara T, Nakagome Y, Ogasawara N, et al (1982) Maternally transmitted extra ring(21) chromosome in a boy with Down's syndrome. Hum Genet 60:78–79

Mayer M, Matsuura J, Jacobs P (1978) Inversions and other unusual heteromorphisms detected by C-banding. Hum Genet 45:43–50

Mikelsaar A-V, Schmid M, Krone W, et al (1977) Frequency of Ag-stained nucleolus organizer regions in the acrocentric chromosomes of man. Hum Genet 37:73–77

Moorhead PS (1976) A closer look at chromosomal inversions. Am J Hum Genet 28:294–296

Nakajima S, Yanagisawa M, Kamoshita S, et al (1976) Mental retardation and congenital malformations associated with a ring chromosome 9. Hum Genet 32:289–293

Niebuhr E (1978a) Cytologic observations in 35 individuals with a 5p− karyotype. Hum Genet 42:143–156

Niebuhr E (1978b) The cri du chat syndrome. Hum Genet 44:227–275

Nielsen KB, Dyggve H, Friedrich U, et al (1978) Small metacentric nonsatellited extra chromosome. Hum Genet 44:59–69

Niss R, Passarge E (1975) Derivative chromosomal structures from a ring chromosome 4. Humangenetik 28:9–23

Opitz JM, Patau K (1975) A partial trisomy 5p syndrome. In: New chromosomal and malformation syndromes. Birth defects: original article series, Vol II. The National Foundation-March of Dimes, New York, pp 191–200

Rhomberg K (1984) Ring chromosome 21 in a healthy woman with three spontaneous abortions. Hum Genet 67:120

Riley SB, Buckton KE, Ratcliffe SG, et al (1981) Inheritance of a ring 14 chromosome. J Med Genet 18:209–213

Rodiere M, Donadio D, Emberger J-M, et al (1977) Isochromosomie 18:46,XX,i(18q). Ann Pediat 24:611–616

Sanchez O, Yunis JJ (1977) New chromosome techniques and their medical applications. In: Yunis JJ (ed) New chromosomal syndromes. Academic, New York, pp 1–54

Sarto GE, Therman E (1980) Replication and inactivation of a dicentric X formed by telomeric fusion. Am J Obstet Gynecol 136:904–911

Schinzel A (1984) Catalogue of unbalanced chromosome aberrations in man. De Gruyter, Berlin

Schmid W, Tenconi R, Baccichetti C, et al (1983) Ring chromosome 21 in phenotypically apparently normal persons: report of two families from Switzerland and Italy. Am J Med Genet 16:323–329

Schnedl W, Abraham R, Dann O, et al (1981) Preferential fluorescent staining of heterochromatic regions in human chromosomes 9, 15, and the Y by D287/170. Hum Genet 59:10–13

Schroeder H-J, Schleiermacher E, Schroeder TM, et al (1967) Zur klinischen Differentialdiagnose des Cri du Chat-Syndroms. Humangenetik 4:294–304

Smith DW (1976) Recognizable patterns of human malformation. 2nd edn. Saunders, Philadelphia

Surana RB, Bailey JD, Conen PE (1971) A ring-4 chromosome in a patient with normal intelligence and short stature. J Med Genet 8:517–521

Tharapel AT, Summitt RL (1978) Minor chromosome variations and selected heteromorphisms in 200 unclassifiable mentally retarded patients and 200 normal controls. Hum Genet 41:121–130

Therkelsen AJ, Hultén M, Jonasson J, et al (1973) Presumptive direct insertion within chromosome 2 in man. Ann Hum Genet 36:367–373

Toomey KE, Mohandas T, Sparkes RS, et al (1978) Segregation of an insertional chromosome rearrangement in 3 generations. J Med Genet 15:382–387

Trunca Doyle C (1976) The cytogenetics of 90 patients with idiopathic mental retardation/malformation syndromes and of 90 normal subjects. Hum Genet 33:131–146

Trunca C, Opitz JM (1977) Pericentric inversion of chromosome 14 and the risk of partial duplication of 14q (14q31→14qter). Am J Med Genet 1:217–228

Uchida IA, McRae KN, Wang HC, et al (1965) Familial short arm deficiency of chromosome 18 concomitant with arhinencephaly and alopecia congenita. Am J Hum Genet 17:410–419

Valcárcel E, Benítez J, Martínez P, et al (1983) Cytogenetic recombinants from a female carrying a paracentric inversion of the short arm of chromosome number 5. Hum Genet 63:78–81

Verma RS, Dosik H (1980) Human chromosomal heteromorphisms: nature and clinical significance. Int Rev Cytol 62:361–383

Vianna-Morgante AM, Nozaki MJ, Ortega CC, et al (1976) Partial monosomy and partial trisomy 18 in two offspring of carrier of pericentric inversion of chromosome 18. J Med Genet 13:366–370

Vogel W, Back E, Imm W (1978) Serial duplication of 10(q11→q22) in a patient with minor congenital malformations. Clin Genet 13:159–163

Wisniewski L, Politis GD, Higgins JV (1978) Partial tetrasomy 9 in a liveborn infant. Clin Genet 14:147–153

Yunis E, Silva R, Egel H, et al (1978) Partial trisomy-5p. Hum Genet 43:231–237

Zackai EH, Breg WR (1973) Ring chromosome 7 with variable phenotypic expression. Cytogenet Cell Genet 12:40–48

XXIII
Reciprocal Translocations

Occurrence

Reciprocal translocations or interchanges have been observed in most organisms, plants as well as animals, that have been studied cytogenetically (cf. Burnham, 1962; White, 1978). They may occur as floating or stable polymorphisms or in single individuals. Just like other chromosome structural changes, such as Robertsonian translocations or inversions, translocations that are originally heterozygous may become homozygous, and this sometimes provides a mechanism for isolating two populations. The spontaneous rate of interchanges is estimated to lie between 10^{-1} and 10^{-3} per gamete per generation in such different organisms as *Drosophila*, grasshopper, mouse, and humans (Lande, 1979).

In plants, balanced carriers of heterozygous translocations are usually discovered because a proportion of the pollen grains are abnormal in shape and size. In animals, balanced carriers are often detected because they are semisterile, and their litter sizes are smaller than normal. In humans, the ascertainment is either through a phenotypically abnormal offspring, infertility, or by chance in cytogenetic surveys of variously chosen populations.

By now translocations involving all chromosome arms have been observed (cf. Borgaonkar, 1984; de Grouchy and Turleau, 1984; Schinzel, 1984). A reciprocal translocation gives rise either to two abnormal but monocentric chromosomes, or to a dicentric and an acentric chromosome. Dicentrics have been discussed in Chapter XXII. Whole-arm transfers or Robertsonian translocations are dealt with in Chapter XXIV. A variety of highly complicated translocations between several chromosomes have also been published, as described later in this chapter.

Breakpoints in Reciprocal Translocations

Human chromosomes do not break at random but mainly in the Q-dark regions. Breaks are also distributed differently with particular chromosome-breaking agents. The location of hot spots depends on the chromosome-breaking agent. Nonrandomness also applies to so-called spontaneous breaks—the ones whose cause we do not know.

The number of breaks that can be analyzed when the material consists of individual cells is naturally of an order of magnitude different from the number of translocation carriers whose chromosomes can be studied. Consequently the fact that the various studies deal with limited material has led to contradictory claims. The confusion is compounded by the pooling of translocations ascertained in different ways. A group ascertained through an unbalanced individual is different from that detected by chance, since the former is affected by powerful selection resulting from the differential viability of individuals partially trisomic or monosomic for various chromosome regions.

The breakpoints in reciprocal translocations seem also to be localized mainly in the Q-dark regions. However, Aurias et al (1978), using several banding techniques on the same translocations, claimed that an excess of breakpoints occurred at the borders of light and dark bands. A different result has been obtained by Nakagome et al (1983) in a careful study of breakpoints in inverted duplications, isodicentrics, and rings in which the sites can be determined more accurately. They concluded that the breaks occurred overwhelmingly in the Q-dark regions and only exceptionally at the borders of bands.

The most extensive analysis so far has been done by C Trunca (Trunca et al, 1981; C Trunca and N Mendel, unpublished) on 863 translocations. When the translocation was ascertained by chance, the breaks occurred at random points. When the ascertainment was through an unbalanced individual, the breakpoints were nonrandom, with an excess of breaks at the telomeres. Moreover, chromosomes 9, 11, 13, 18, 21, and 22 were significantly overrepresented, whereas chromosomes 1, 2, 3, 6, 7, and 19 were significantly underrepresented.

A related question is whether chromosomes exchange segments at random or whether specific combinations are preferred. Apart from a couple of exceptions, the exchanges seem to be at random (C Trunca and N Mendel, unpublished). The most glaring exception is the affinity of the segments 11q23 and 22q11. Of some 100 translocations involving 11q, in only 13 cases was the other chromosome not 22 (Fraccaro et al, 1980; Zackai and Emanuel, 1980; Pihko et al, 1981). This suggests that the two bands involved in 11q and 22q have a homologous sequence which has a tendency to pair and cross over. This hypothesis is further borne out by the similarity of symptoms caused by partial trisomy for the distal segment of 11q and for 22q. Other combinations that occur significantly more frequently than expected are t(9;22) and t(9;15) (C Trunca and N Mendel, unpublished).

Multiple Rearrangements

In many organisms, including humans, individuals with complicated chromosome rearrangements have been found. Thus Palmer et al (1976) described in the mother of a chromosomally unbalanced child a rearrangement involving chromosomes 3, 11, and 20. They also listed 14 previously published multiple-break cases. Bijlsma et al (1978), who described an exceptional family with two reciprocal translocations in three generations, leading to the birth of one unbalanced individual, mention a few more complex rearrangements. Later cases have been collected by Meer et al (1981), who described the segregation of a complex rearrangement between chromosomes 6, 7, 8, and 12 through three generations.

A phenotypically abnormal child with five structurally aberrant chromosomes 1, 4, 7, 12, and 15 was born to a woman who, during pregnancy, developed malignant melanoma, which was not treated before the child was born (Fitzgerald et al, 1977). One cannot help wondering whether the same unknown agent might have been responsible for the malignant disease in the mother and the chromosome aberrations in the child. The record in multiple breaks may have been reached by two boys, one normal, one abnormal, who each had six chromosomes involved in abnormalities (Prieto et al, 1978; Watt and Couzin, 1983).

Based on present evidence, it is impossible to decide whether such complex chromosome rearrangements, although rare, might still be more common than expected if breaks occurred independently. Obviously publication bias is a significant factor, since the more complicated a chromosome rearrangement is, the more certain it is to be published. However, it is also true that many agents, such as cosmic rays, viruses, or mutagenic substances, affect cells nonrandomly.

Phenotypes of Balanced Translocation Carriers

There is no doubt that the overwhelming majority of persons with a balanced reciprocal translocation are phenotypically normal. However, some observations indicate that at least certain reciprocal translocations may affect the phenotype. This evidence comes from studies in which the frequency of apparently balanced reciprocal translocations is compared in mentally retarded and control populations. For instance, Funderburk et al (1977) found in 455 retarded children seven reciprocal translocations, whereas the corresponding number in 1679 nonretarded children with psychiatric problems was four (P<0.05). By pooling the results of various surveys on mental retardates, the authors conclude that these patients have five times more balanced, mainly de novo, reciprocal translocations than consecutive newborns. Interestingly, the incidence of balanced Robertsonian translocations does not seem to be increased in mentally retarded patients. Whether breaks in particular chromosome regions are more likely to have phenotypic effects, as suggested by Biederman and Bowen (1976) for 12q12-q21, is not clear at present. Although rare, the possibility of phenotypic effects should not be neglected in genetic counseling when amniocentesis reveals an apparently balanced reciprocal translocation.

Another expression of a position effect is the male sterility caused by reciprocal translocations between autosomes both in the mouse (Searle, 1982) and in humans (Zuffardi and Tiepolo, 1982). In infertile men the incidence of balanced non-Robertsonian translocations is increased some four times as compared with the newborn (Zuffardi and Tiepolo, 1982). However, the first position effect described in humans concerned gonadal dysgenesis in the female. In a woman the "critical region" on the long arm of both X chromosomes must be intact to allow normal sex development (Sarto et al, 1973; Therman, 1983) (Chapter XX). Position effect seems also to play an important role in the activation of oncogenes that are placed in new positions through reciprocal translocations (Chapter XXVI).

Although position effect has been suggested in many plants and animals, few proven cases exist. Obviously there are different kinds of position effect. The best known is the so-called variegation type, which has been studied in *Drosophila* and the mouse. A variegated phenotype is caused by a gene being transferred next to heterochromatin; this results in the gene's being active in some cells but not in others. This kind of position effect has not been observed in man.

Several hypotheses have been put forward to explain the phenotypic effect of some of the apparently balanced reciprocal translocations described above (cf. Hecht et al, 1978). One is that a gene functions differently at a new location. Or the break may occur within a gene, thus destroying its function, allowing a recessive allele to express itself. This is the most attractive explanation for the cases in which a normal parent has an abnormal child, while both show the same translocation. The possibility that a minute deletion or duplication has arisen at the break site cannot be excluded, although a high-resolution study of several balanced translocations in abnormal carriers failed to reveal any imbalance (Raimondi et al, 1983).

Phenotypes of Unbalanced Translocation Carriers

An unbalanced translocation involves partial trisomy for one chromosome and partial monosomy for another, although the deleted segment may be very small or almost nonexistent if one of the breaks lies at a telomere. Trisomy and monosomy effects practically always include mental retardation and multiple congenital anomalies. The analysis of the symptoms is often made difficult by simultaneous monosomy and trisomy. By now the chromosomes and phenotypes of thousands of unbalanced translocation carriers have been described. The reader is referred to the compendiums of Borgaonkar (1984), de Grouchy and Turleau (1984), and Schinzel (1984). Cytogenetic surveys on variously ascertained mental retardates show that chromosome abnormalities constitute an important cause of mental retardation and congenital anomalies. For instance, in the Madison Blind Study, the chromosomes of 410 mentally retarded patients with congenital anomalies of unknown etiology were compared with the same number of normal controls. About 6 percent of the patients showed a structurally abnormal, unbalanced chromosome constitution, whereas none of the controls did (C Trunca, unpublished).

Jacobs et al (1978) summarized the results of a number of cytological studies on mentally retarded populations. If the patients were unselected, 2 percent had a chromosome anomaly, excluding those with Down's syndrome. In selected patients, which were generally more retarded, 11 percent showed an abnormal chromosome constitution. If one includes only the autosomal trisomies—again excluding 21-trisomy—and the structurally unbalanced chromosome constitutions, the corresponding number is 8.4 percent.

Even if breaks occurred exclusively in Q-dark bands, the number of random interchanges would be enormous. However, the range of combinations that allow the birth of a live child is much more limited, since most sizable deletions are lethal, as are many duplications. This strong selection has led to a relatively high incidence of certain partial trisomies and monosomies, whereas others are extremely rare or nonexistent, which is obvious from any survey of the literature. The following parameters probably determine this unequal distribution: (1) the length of the segment involved; (2) the Q-darkness versus brightness of the segment; the Q-darker a region is, the more severe are the effects of imbalance; (3) individual genes that may act as trisomy or monosomy lethals; and (4) special "hot spots," short Q-dark bands, which have been assumed to contain a high density of genes, and which may also act as trisomy and monosomy lethals (Korenberg et al, 1978; Kuhn et al, 1985).

The observations made on partial 11q trisomy (Pihko et al, 1981) serve as an illustration of the last point. In 20 cases in which the trisomy ranged from most of 11q to 11q23→ter11q, there was no difference in the quality or quantity of the symptoms. The inevitable conclusion is that most—possibly all—of the symptoms are caused by trisomy for the short segment distal to 11q23. Similar studies on partial trisomies and monosomies for other chromosome arms would be of interest.

Minute duplications and deletions have been discussed in Chapter XXII. In Fig. XXIII.1, the relevant chromosomes of a family with a very small reciprocal translocation 3p;21p are illustrated diagrammatically (A

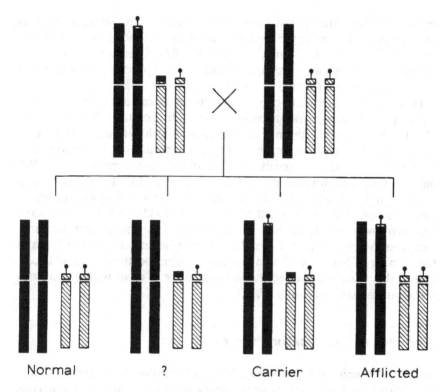

Normal   ?   Carrier   Afflicted

Figure XXIII.1. Diagram of the chromosomes of a family with t(3p;21p) (the diagram was drawn before the identity of the acrocentric was known). It is not known whether individuals with partial trisomy for 3p exist (courtesy of A Drewry).

Drewry, unpublished). Indeed, this exchange is so small that, but for the satellites on 3p, it would never have been discovered. The family had three carriers and four affected members in three generations. The symptoms that were caused by the minute deletion of 3p included mental retardation, a sacral dimple, and abnormal ears and teeth (A Drewry, unpublished). In the carriers, the chromosomes 3 and 21 obviously form two bivalents, resulting for once in 1:1:1:1 (normal:carrier:one type of unbalanced:another type of unbalanced segregation). However, a family member with partial trisomy for 3p could not be cytologically ascertained, and such a minute duplication probably would not cause any recognizable symptoms.

Fetal Death

The lethal effects of the various trisomies and/or monosomies are reflected in the increased rate of spontaneous abortions and stillbirths in many translocation families. Often a very early abortion remains unrecognized. Thus a sizable translocation 3q−;4p+ in the father apparently resulted in sterility after the birth of a normal carrier son (Sarto and Therman, 1976). In another family, Nuzzo et al (1973) found a maternal translocation in which the entire 2q was attached to 1p, resulting in four recognized abortions and no live births.

However, the best data come from chromosome analyses of couples with repeated spontaneous abortions. Depending on the selection of couples for such studies (the number of abortions, habitual abortion combined with abnormal offspring, whether other causes for abortion have been ruled out, etc.), the results range from 0 to 31 percent of the couples having reciprocal translocations (cf. Ward et al, 1980). In a review of studies on 1,331 couples, the incidence of Robertsonian and balanced translocations was 6.2 percent (Davis et al, 1982). Other studies have reported lower estimates, such as 3.6 percent of the couples having translocations (Michels et al, 1982) or 2.2 percent (Pantzar et al, 1984). Usually the number of Robertsonian and reciprocal translocations is about equal, and to this should be added the same number of cases with X chromosome mosaicism (including 47,XXX).

Examples of Translocation Families

Families in which one parent is a balanced translocation carrier fall into the following classes: (1) those in which none of the possible abnormal offspring is viable; (2) those in which one type of offspring, usually the one with the smaller deletion, is born alive; and (3) those in which two types of abnormal offspring are viable. Only a few rather arbitrarily chosen translocation families are presented here as examples.

Families in which all pregnancies end up as spontaneous abortions have been mentioned in the previous section (for example, Nuzzo et al, 1973).

A typical representative of the second group was studied by Pihko et al (1981) (Fig. XXIII.2). The abnormal daughter was severely retarded mentally and showed diverse congenital anomalies. Chromosome analysis revealed a too long 4q, whereas the mother and sister were balanced carriers of t(11q−; 4q+). The spontaneous abortions were not available for cytogenetic studies, but it is possible that they represented other types of unbalanced chromosome constitutions. Since the segment from 11q is apparently attached to the very end of 4q, the symptoms of the proposita are probably caused mainly by trisomy for about one-half of 11q. Another similar family has been presented by Centerwall et al (1976); here the segregation of t(9p−;14q+) in four generations resulted in six affected individuals with partial trisomy 9p and four phenotypically normal carriers. Another typical result of a reciprocal translocation 7q−;21q+ in a male was the birth of an abnormal daughter trisomic for the end region

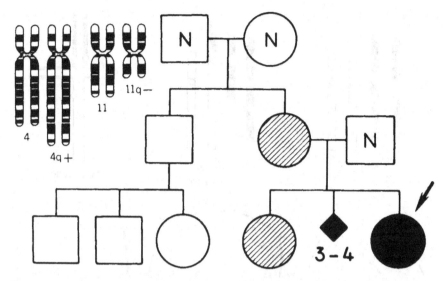

Figure XXIII.2. A family in which the mother and one daughter (shaded) were balanced carriers of t(4q+;11q−), one daughter (black) had the unbalanced translocation chromosome 4q+, and abortions possibly had 11q− in unbalanced state; N chromosomes normal (Pihko et al, 1981).

of 7q and a spontaneous abortion whose chromosome displayed the corresponding deletion (Bass et al, 1973).

Fortunately it is rare that more than one type of unbalanced offspring is born alive in the same translocation family. The occurrence of cri du chat syndrome and its trisomic "countertype" is discussed in a previous chapter. For a number of other translocations, a similar segregation has been described. To take a random example, Jacobsen et al (1973) analyzed three generations of a family in which 14 normal carriers had t(11q−;21q+), two abnormal segregants showed partial monosomy for 11q, and one had trisomy for the same segment. As might be expected, the monosomic individuals were more severely affected than was the trisomic one.

A 1:3 segregation (one of the four chromosomes goes to one pole while three go to the other) in a translocation carrier gives rise to unbalanced offspring with 45 or 47 chromosomes. Such disjunction is discussed in more detail in connection with the risk of abnormal offspring in translocation carriers. Eight kinds of offspring are possible from such a segregation. They include monosomy or trisomy for a normal chromosome. An extra translocated chromosome combined with an otherwise normal complement (47 chromosomes) has been termed *tertiary trisomy*, and a 45-chromosome complement that includes an abnormal chromosome is *tertiary monosomy*.

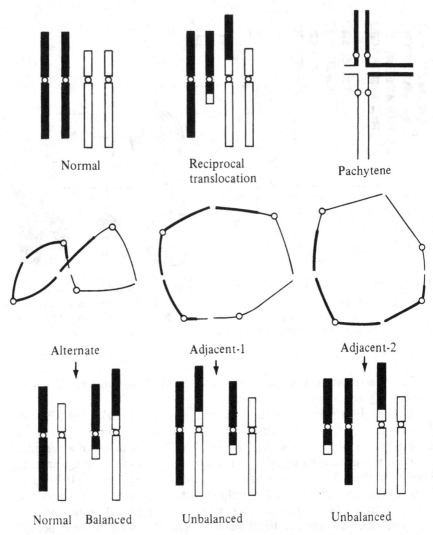

Figure XXIII.3. Reciprocal translocation between two chromosomes, pachytene configuration of the four chromosomes, and modes of orientation of a ring of four in metaphase I: alternate, adjacent–1, adjacent–2 (1:3 segregation not shown). The gametes formed are: normal, balanced, two types of unbalanced (zygotes possibly viable), and two types of even more unbalanced (zygotes presumably nonviable). In adjacent-1 segregation homologous centromeres separate; in adjacent-2 segregation homologous centromeres go to the same pole.

Interchange trisomy implies that the two translocated chromosomes are present together with an extra normal chromosome (47 chromosomes) (cf. Lindenbaum and Bobrow, 1975). A rare situation in which two different types of abnormal offspring resulted from 1:3 segregation is represented by a woman with t(9q−;21p+) who gave birth to one daughter with an extra 9p chromosome and another who was 21-trisomic, in addition to a chromosomally normal daughter (Habedank and Faust, 1978). Her translocation had features that are known to promote this type of disjunction, namely, the chromosomes were of different size, one was acrocentric, and the breaks were near the centromere. Another reciprocal translocation which agrees with these rules and practically always segregates 1:3 is t(11;22)(q23;q22). In 72 families the mother was the carrier, in two the father (cf. Fraccaro et al, 1980; Zackai and Emanuel, 1980). Normals and balanced carriers were equally frequent in the offspring.

In an interesting family with five abortions, three were studied cytologically by Kajii et al (1974). Segregation in the father, who had a t(13q−;18q+), led to three chromosomally analyzed abortions; 47,+13 (tertiary trisomy), 46,13q− (unbalanced translocation), and 47,t(13q−; 18q+)+18 (interchange trisomy). The abortion with 46 chromosomes resulted from an adjacent−1 2:2 disjunction (Fig. XXIII.3), the other two from a 1:3 segregation.

Meiosis in Translocation Carriers

Since homologous chromosome segments tend to pair in meiosis even when one of them is translocated to another chromosome, the two normal and the two translocation chromosomes often form a group of four, which in some organisms can be seen as a cross-like figure in pachytene (Fig. XXIII.3). Very small translocated segments may remain unpaired or fail to form a chiasma, which results in the formation of two bivalents or a group of three and a univalent. However, if at least one chiasma is present in each arm of the pachytene cross, the metaphase configuration will be a ring of four (Fig. XXIII.3); or if a chiasma fails to form in one arm, the result will be a chain. A ring or chain of four may orient itself in different ways on the spindle.

An alternate orientation (Fig. XXIII.3) gives rise to one cell with normal chromosomes and another with the translocation chromosomes. Adjacent−1 orientation (Fig. XXIII.3) leads to the formation of two unbalanced cells, each with one translocation chromosome. Adjacent−2 orientation (Fig. XXIII.3) gives rise to two cells that are usually even more unbalanced. Various types of 1:3 segregation are also possible, as discussed above. Sometimes two bivalents are formed, and their independent segregation produces equal numbers of normal, carrier, and two types of unbalanced gametes (Fig. XXIII.1). It should be stressed that there is *no theoretical expectation* for the segregation ratios of human reciprocal translocations, since the relative frequencies of the different orientations of a specific ring or chain are unknown, so we have to rely on empirical risk figures. Usually, however, complementary classes from the same orientation are equally frequent.

Genetic Risk for Translocation Carriers

One of the most important unbiased sources of reciprocal translocations is provided by studies of the newborn (cf. Hook and Hamerton, 1977). In 56,952 infants, 51 balanced reciprocal and insertional translocations were found, which happens to be the number of balanced Robertsonian translocations in the same material (Table XXIII.1). This amounts to one balanced translocation per 1000 neonates. A much smaller number of unbalanced translocations or insertions, that is, 7 (0.1/1000), was encountered in the same newborn population. Other findings of this study are given in Table XXIII.1.

It has been obvious for a long time that the risk figures for balanced translocation carriers having a chromosomally abnormal offspring are very different for translocations ascertained by chance than for those found through an unbalanced offspring. C Trunca and N Mendel (unpublished), in their analysis of 863 translocations, give the following figures. In the unbiased group the risk for a chromosomally unbalanced child was 1.5 percent and that for miscarriage or stillbirth 25.0 percent. On the other hand, when the reciprocal translocation was ascertained through a chromosomally unbalanced child, the risk for chromosomally unbalanced offspring was 16.2 percent and for a miscarriage or stillbirth 25.1 percent.

Table XXIII.1. Chromosome Abnormalities in 56,952 Newborn Infants.[a]

Abnormality	Incidence (per 1000 neonates)
Aneuploidy (including mosaicism)	
Sex chromosomes[b]	
Males	2.61
Females	1.51
Autosomes	1.44
Structural abnormalities	
Balanced Robertsonian translocations	0.90
Balanced translocations	0.90
Unbalanced Robertsonian translocations	0.07
Unbalanced translocations	0.12
Other structural abnormalities	0.55
Total	6.05

[a]Modified from Hook and Hamerton, 1977.
[b]Numbers apply to infants of the relevant sex.

Whatever the ascertainment of the reciprocal translocation, the ratio of normal to balanced carrier offspring is about 1:1 (C Trunca and N Mendel, unpublished), since these are complementary and usually both are normal.

The risk figures for different types of translocations depend on the pachytene configuration which the four chromosomes form. This again is determined by the shape of the chromosomes, the distance of the breakpoints from the centromere (interstitial segment), and from the telomeres, which in turn affect the probability of crossing-over in the exchanged and the interstitial segments. The orientation of a translocation quadrivalent has been directly studied in the meiosis of many organisms, but not in humans (cf. Lewis and John, 1963).

In humans the orientation can be concluded from the empirical risk figures obtained from a large number of translocation families (C Trunca and N Mendel, unpublished). Now the meiotic segregation can also to some extent be determined by analyzing the human chromosomes in the male pronucleus in a hamster egg fertilized by a human sperm (Balkan and Martin, 1983).

According to C Trunca and N Mendel (unpublished) the types of segregation taking place in the variously constructed pachytene configurations are the following: (1) If the interstitial segments are so short that they do not form chiasmata, of 10 abnormal children, 6 were the result of adjacent-1 and 4 of adjacent-2 segregation. (2) In the families in which at least one interstitial segment was expected to have a chiasma, 390 of 395 abnormal offspring resulted from an adjacent-1 segregation. (3) If the translocated segments are very short, the four chromosomes will form a chain or two bivalents. In such families, of 530 unbalanced offspring, 525 were the result of an adjacent-1 segregation. (4) If the non-exchange part is very short, thus being unable to form a chiasma, the resulting chain is expected to undergo an adjacent-2 segregation and, indeed, 30 of 36 children in this group came from this type of segregation. (5) The following factors seem to favor 1:3 over 2:2 segregation: one of the chromosomes is acrocentric, at least one break is near the centromere, and the participating chromosomes are very unequal in size; in other words, the pachytene configuration is highly asymmetrical (Lindenbaum and Bobrow, 1975; C Trunca and N Mendel, unpublished). In this group 77 chromosomally abnormal children were found, of which 50 had 45 or 47 chromosomes. When the pachytene cross was relatively symmetrical, only 19.8 percent of the chromosomally abnormal children resulted from a 1:3 segregation. The translocations for which 1:3 segregation has been observed involve the chromosomes very nonrandomly. Thus interchange trisomy has only been found for chromosomes 18, 21, and 22, whereas 9p seems to be preferred in tertiary trisomy (Lindenbaum and Bobrow, 1975).

About 90 percent of chromosomally unbalanced children resulting from 1:3 disjunction have been born to carrier mothers. The risks for further unbalanced live-born offspring when the family has been ascertained through a 1:3 segregant are in the range of 10 percent to 20 percent for a carrier mother and 0 for the father (Lindenbaum and Bobrow, 1975).

The overall risks of abnormal offspring for male and female carriers are different (C Trunca and N Mendel, unpublished). For males (996 carriers) the risk was 6.6 percent and for females (1539 carriers) 11.9 percent.

The highest and the lowest risks are found in the following translocation groups (C Trunca and N Mendel, unpublished). If the exchanged segments are so short that crossing-over rarely takes place in either side arm of a pachytene cross, a chain or two bivalents are formed, and the risk for a live-born abnormal child is 21.0 percent and that of a miscarriage or stillborn 25.7 percent. The low risk group consists of the translocations in which the exchanged segments are relatively long, whereas the interstitial segments are short, which leads to the formation of a ring. The risk of a live-born, chromosomally abnormal child is 2.5 percent and that of miscarriage or stillbirth 33.7 percent.

The risk of abnormal offspring in the high risk group was 17 percent for male carriers and 19.7 percent for females. In the low risk group the corresponding numbers were 0 percent for the males and 4 percent for the females. The risk for stillbirths and miscarriages in the high risk group was 28.9 percent for the males and 24.5 percent for the females, whereas the corresponding numbers in the low risk group were 28.3 percent for the males and 40.2 percent for the females (C Trunca and N Mendel, unpublished).

References

Aurias A, Prieur M, Dutrillaux B, et al (1978) Systematic analysis of 95 reciprocal translocations of autosomes. Hum Genet 45:259–282

Balkan W, Martin RH (1983) Chromosome segregation into the spermatozoa of two men heterozygous for different reciprocal translocations. Hum Genet 63:345–348

Bass HN, Crandall BF, Marcy SM (1973) Two different chromosome abnormalities resulting from a translocation carrier father. J Pediatr 83:1034–1038

Biederman B, Bowen P (1976) Balanced translocations involving chromosome 12: report of a case and possible evidence for position effect. Ann Génét 19:257–260

Bijlsma JB, France HF de, Bleeker-Wagemakers LM, et al (1978) Double translocation t(7;12),t(2;6) heterozygosity in one family. Hum Genet 40:135–147

Borgaonkar DS (1984) Chromosomal variation in man: a catalog of chromosomal variants and anomalies, 4th edn. Liss, New York

Burnham CR (1962) Discussions in cytogenetics. Burgess, Minneapolis

Centerwall WR, Miller KS, Reeves LM (1976) Familial "partial 9p" trisomy: six cases and four carriers in three generations. J Med Genet 13:57–61

Davis JR, Weinstein L, Veomett IC, et al (1982) Balanced translocation karyotypes in patients with repetitive abortion. Am J Obstet Gynecol 144:229–233

Fitzgerald PH, Miethke P, Caseley RT (1977) Major karyotypic abnormality in a child born to a woman with untreated malignant melanoma. Clin Genet 12:155–161

Fraccaro M, Lindsten J, Ford CE, et al (1980) The 11q;22q translocation: a European collaborative analysis of 43 cases. Hum Genet 56:21–51

Funderburk SJ, Spence MA, Sparkes RS (1977) Mental retardation associated with "balanced" chromosome rearrangements. Am J Hum Genet 29:136–141

Grouchy J de, Turleau C (1984) Clinical atlas of human chromosomes, 2nd edn. Wiley, New York

Habedank M, Faust J (1978) Trisomy 9p and unusual translocation mongolism in siblings due to different 3:1 segregations of maternal translocation rcp(9;21)(p11;q11). Hum Genet 42:251–256

Hecht F, Kaiser-McCaw B, Patil S, et al (1978) Are balanced translocations really balanced? Preliminary cytogenetic evidence for position effect in man. In: Summitt RL, Bergsma D (eds) Sex differentiation and chromosomal abnormalities. Birth defects: original article series, Vol 14, No. 6C. The National Foundation-March of Dimes, New York, pp 218- 286

Hook EB, Hamerton JL (1977) The frequency of chromosome abnormalities detected in consecutive newborn studies—differences between studies— results by sex and by severity of phenotypic involvement. In: Hook EB, Porter IH (eds) Population cytogenetics. Academic, New York, pp 63–79

Jacobs PA, Matsuura JS, Mayer M, et al (1978) A cytogenetic survey of an institution for the mentally retarded: I. Chromosome abnormalities. Clin Genet 13:37–60

Jacobsen P, Hauge M, Henningsen K, et al (1973) An (11;21) translocation in four generations with chromosome 11 abnormalities in the offspring. Hum Hered 22:568–585

Kajii T, Meylan J, Mikamo K (1974) Chromosome anomalies in three successive abortuses due to paternal translocation, t(13q−;18q+). Cytogenet Cell Genet 13:426–436

Korenberg JR, Therman E, Denniston C (1978) Hot spots and functional organization of human chromosomes. Hum Genet 43:13–22

Kuhn EM, Therman E, Denniston C (1985) Mitotic chiasmata, gene density, and oncogenes. Hum Genet 70:1–5

Lande R (1979) Effective deme sizes during long-term evolution estimated from rates of chromosomal rearrangement. Evolution 33:234–251

Lewis KR, John B (1963) Spontaneous interchange in Chorthippus brunneus. Chromosoma 14:618–637

Lindenbaum RH, Bobrow M (1975) Reciprocal translocations in man. 3:1 meiotic disjunction resulting in 47- or 45-chromosome offspring. J Med Genet 12:29–43

Meer B, Wolff G, Back E (1981) Segregation of a complex rearrangement of chromosomes 6, 7, 8 and 12 through three generations. Hum Genet 58:221–225

Michels VV, Medrano C, Venne VL, et al (1982) Chromosome translocations in couples with multiple spontaneous abortions. Am J Hum Genet 34:507–513

Nakagome Y, Matsubara T, Fujita H (1983) Distribution of break points in human structural rearrangements. Am J Hum Genet 35:288–300

Nuzzo F, Giorgi R, Zuffardi O, et al (1973) Translocation t(1p+;2q−) associated with recurrent abortion. Ann Génét 16:211–214

Palmer CG, Poland C, Reed T, et al (1976) Partial trisomy 11, 46,XX,−3,−20,+ der3,+der20,t(3:11:20), resulting from a complex maternal rearrangement of chromosomes 3, 11, 20. Hum Genet 31:219–225

Pantzar JT, Allanson JE, Kalousek DK, et al (1984) Cytogenetic findings in 318 couples with repeated spontaneous abortion: a review of experience in British Columbia. Am J Med Genet 17:615–620

Pihko H, Therman E, Uchida IA (1981) Partial 11q trisomy syndrome. Hum Genet 58:129–134

Prieto F, Badia L, Moreno JA, et al (1978) 10p– syndrome associated with multiple chromosomal abnormalities. Hum Genet 45:229–235

Raimondi SC, Luthardt FW, Summitt RL, et al (1983) High-resolution chromosome analysis of phenotypically abnormal patients with apparently balanced structural rearrangements. Hum Genet 63:310–314

Sarto GE, Therman E (1976) Large translocation t(3q–;4p+) as probable cause for semisterility. Fert Steril 27:784–788

Sarto GE, Therman E, Patau K (1973) X inactivation in man: a woman with t(Xq–;12q+). Am J Hum Genet 25:262–270

Schinzel A (1984) Catalogue of unbalanced chromosome aberrations in man. De Gruyter, Berlin

Searle AG (1982) The genetics of sterility in the mouse. In: Crosignani PG, Rubin BL (eds) Genetic control of gamete production and function. Academic, London, pp 93–114

Therman E (1983) Mechanisms through which abnormal X-chromosome constitutions affect the phenotype. In: Sandberg AA (ed) Cytogenetics of the mammalian X chromosome, Part B: X chromosome anomalies and their clinical manifestations. Liss, New York, pp 159–173

Trunca C, Weiner D, Kaplan A (1981) The meiotic behavior of reciprocal translocations. Am J Hum Genet 33: 124A

Ward BE, Henry GP, Robinson A (1980) Cytogenetic studies in 100 couples with recurrent spontaneous abortions. Am J Hum Genet 32:549–554

Watt JL, Couzin DA (1983) De novo translocation heterozygote with three reciprocal translocations. J Med Genet 20:385–388

White MJD (1978) Modes of speciation. Freeman, San Francisco

Zackai EH, Emanuel BS (1980) Site-specific reciprocal translocation, t(11;22)(q23;q11), in several unrelated families with 3:1 meiotic disjunction. Am J Med Genet 7:507–521

Zuffardi O, Tiepolo L (1982) Frequencies and types of chromosome abnormalities associated with human male infertility. In: Crosignani PG, Rubin BL (eds) Genetic control of gamete production and function. Academic, London, pp 261–273

XXIV
Robertsonian Translocations

Occurrence

Robertsonian translocations refer to the recombination of whole chromosome arms. Such translocations take place most often between acrocentric or telocentric chromosomes. They have played an important role in the evolution of both plants and animals, as demonstrated by organisms within a species (or in closely related species) that have different chromosome numbers but the same number of chromosome arms. In the humans, Robertsonian translocations are the most common structurally abnormal chromosomes. Surprisingly, they seem to be much rarer in some other animal species, such as mice or cattle (cf. Ford, 1970).

Whole-arm transfers between human chromosomes other than the acrocentrics seem to be extremely rare; only a couple of cases have been reported. One reason for this scarcity may be that such translocations are not ascertained through abnormal offspring, since monosomy for one chromosome arm combined with trisomy for another would be lethal (cf. Schober and Fonatsch, 1978; Niikawa and Ishikawa, 1983). However, whole-arm transfers between nonacrocentric chromosomes are not found in unselected populations either. In the following discussion the term Robertsonian translocation will be used to describe a whole-arm transfer between acrocentrics.

Robertsonian translocations do not seem to affect the phenotype of a balanced carrier, apart from occasional male sterility. Individuals with 45 chromosomes including a Robertsonian translocation between two long arms are called balanced. That the deletion of the short arms of the acrocentrics does not have any damaging effects is a further indication of their total heterochromasy.

The relatively high frequency of Robertsonian translocations probably reflects their high incidence. On the other hand, the probability of their ascertainment is increased by their familial occurrence. For instance, 85 percent to 95 percent of DqDq translocations are familial in unselected material (cf. Nielsen and Rasmussen, 1976).

The different modes of formation of Robertsonian translocations between two acrocentric chromosomes are shown diagrammatically in Fig. VII.5. One chromosome may break through the short arm and the other through the long arm near the centromere, or both may break through the centromere. In both cases the result is one long and one short monocentric chromosome. If both chromosomes have a break in the short arm, one dicentric and one acentric chromosome are formed. One or both acrocentrics may also break through the satellite stalk, in which case the translocation chromosome has a nucleolar organizing region in the middle (cf. Mikkelsen et al, 1980). The mechanics of the formation of Robertsonian translocations are discussed, for instance, by John and Freeman (1975).

Monocentric and Dicentric Chromosomes

Banding techniques demonstrate unequivocally that both monocentric (Fig. XXIV.1) and dicentric Robertsonian translocations exist (Niebuhr, 1972). However, in individual cases it is often difficult to distinguish between the two. This is not made easier by the fact that one centromere in a dicentric is often inactivated, which means that its site is not marked by a constriction (Niebuhr, 1972). However, even if both centromeres are functioning this does not lead to aberrations in mitosis, since there is almost never a twist between them.

The small chromosome consisting of the short arms of two acrocentrics is usually, but not invariably, lost. The occurrence of such small bisatellited chromosomes, in addition to the normal chromosome complement, was mentioned in a preceding chapter. Among 11,148 Danish newborn infants, six with such chromosomes were found, giving a frequency of 0.54/1000 (Nielsen and Rasmussen, 1975). Palmer et al (1969) have described a rare family with both a DqDq and a DpDp chromosome. Four family members had both, whereas two showed only the long chromosome.

Isochromosomes resulting from misdivision of the centromere cannot be distinguished from Robertsonian translocations between two homologous chromosomes on morphological grounds. However, the history of the chromosome often allows a determination to be made. If a person has one normal cell line and another with 46 chromosomes, including one free 21 and a metacentric consisting of two chromosomes 21, the latter is obviously an isochromosome. On the other hand, in a mosaic with a normal cell line and another with 45,t(21q21q), a Robertsonian translocation

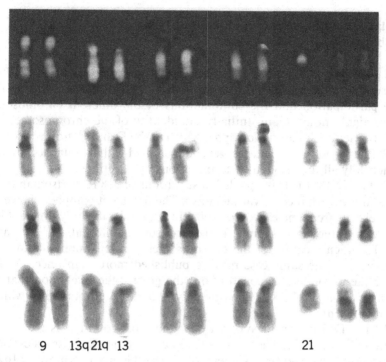

9   13q 21q 13   21

Figure XXIV.1. Partial karyotypes of a woman with 45,XX,t(13q21q), representing a family with many carriers and several Down's syndrome children. One chromosome 9 has a large, partially inverted C-band. The translocation chromosome 13q21q has one short C-band and is monocentric (Sarto and Therman, unpublished).

can be taken for granted. In a family studied in our laboratory, in which three children had Down's syndrome, the father was such a mosaic and the affected children had 46,t(21q21q). Mosaics with a telocentric and an isochromosome for a Dq or a Gq, which arise from each other through misdivision of the centromere, have been discussed in Chapter XVI.

Even more complicated mosaics involving Robertsonian translocations have been encountered. Thus, in two children with features of Down's syndrome, one cell line had 45 chromosomes with a 15q21q translocation, whereas the other showed 46 chromosomes, including a 21q21q chromosome (Atkins and Bartsocas, 1974; Vianna-Morgante and Nunesmaia, 1978). A few individuals with two Robertsonian translocations have been described (for example, Orye and Delire, 1967; Cohen et al, 1968). Furthermore, a therapeutically aborted fetus had 44 chromosomes including two t(14q21q) chromosomes (Rockman-Greenberg et al, 1982). Although 45-chromosome carriers of Robertsonian translocations, as a rule, are normal, the deletion of four nucleolar organizing regions may be deleterious when the incidence of transcribed regions varies between six and ten in normal persons (cf. Mikelsaar et al, 1977).

Relative Frequencies of the Different Types of
Robertsonian Translocations

All possible combinations of acrocentric chromosomes are found in Robertsonian translocations. It became clear at an early stage, however, that the participation and the combinations of the different chromosomes were highly nonrandom. Initially the identity of the chromosomes was determined with autoradiography, which in the D and G groups works so well that when the same cases were later studied with banding techniques, practically all the previous determinations were verified.

Table XXIV.1 lists the pooled observations of 918 Robertsonian translocations ascertained in various ways. The list is not complete; however, the relative frequencies would probably not be significantly different in a more comprehensive sample. Three error sources affect all such attempts: (1) The identification of the chromosomes is often unreliable in early studies. (2) The same case may be published more than once. (3) The rarer the chromosome constitution, the greater the likelihood that the case will be published; consequently the rare combinations will be overrepresented.

Of the DqDq translocations, by far the most common is the combination of 13 and 14. The same position among the DqGq translocations is occupied by 14q21q, whereas of the GqGq translocations 21q21q seems to be more common than 21q22q.

Naturally the mode of ascertainment of Robertsonian translocations has an effect on the figures obtained. Most of the published cases come from three sources: (1) unbiased studies, such as consecutive newborns or chance findings in populations selected for various reasons; (2) studies on families ascertained through an unbalanced individual; (3) studies on persons with infertility problems, including habitual abortions.

Table XXIV.1. Participation of the Different
Acrocentric Chromosomes in 918 Robertsonian
Translocations Ascertained in Various Ways (when
Several Family Members Have the Same
Translocation, It Is Counted as One).[a]

		13	14	15	21	22
	13	31				
	14	251	4			
Chromosome	15	14	19	17		
	21	30	320	38	149	
	22	8	4	5	19	9

[a]Data from 177 studies.

Studies on the Newborn

Hook and Hamerton (1977) have reviewed six studies in which the chromosome constitutions of 56,952 newborn infants were determined. In this material, 51 Robertsonian translocations were found, of which 40 were DqDq and 11 were DqGq combinations. The incidence of Robertsonian translocations in an unselected population is thus 0.09 percent or about 1/1000 (Table XXIII.1). In 21 of the DqDq translocations, the participating chromosomes have been identified; in 20 cases they were 13 and 14, and in one 14 and 15. Four of the identified DqGq translocations involved 14 and 21, one was 13q21q, and one 15q22q (cf. Jacobs, 1977).

Ascertainment Through an Unbalanced Individual

A common mode of ascertainment for Robertsonian translocations is through an individual with Down's syndrome. In patients selected for Down's phenotype, the frequency of Robertsonian translocations ranges from 3.2 percent (Hongell et al, 1972) to 5 percent in 4330 patients (cf. Chapman et al, 1973). The mother's age is also a factor; about 8 percent of Down's syndrome patients whose mothers were under 30 years of age had translocations, whereas if the mothers were over 30, only 1.5 percent of the affected children were translocation cases (cf. Mikkelsen, 1971).

Mikkelsen (1971, 1973) lists 76 DqGq translocations from her laboratory and the literature, ascertained through a Down's syndrome patient. One of them had a t(13q21q), 65 had t(14q21q), 9 had 15q21q, and in one case the chromosome involved was either 14 or 15. Translocations between chromosomes 21 and 22 are often found through an affected person, and translocations between two chromosomes 21 are practically always found this way. Of 12 GqGq chromosomes ascertained in this way, 3 were 21q22q and 9 were 21q21q (Mikkelsen, 1973).

Translocations between chromosomes 13 and 14 are rarely discovered through a 13 trisomic individual, whereas this is the usual ascertainment for 13q13q translocations. Only occasionally has a DqGq chromosome been found through a 13 trisomic individual; one family with a t(13q21q) (Pérez-Castillo and Abrisqueta, 1978) and two families with 13q22q (Abe et al, 1975; Daniel and Lam-Po-Tang, 1976) have been detected in this way.

Ascertainment Through Infertility

Both sterility and habitual abortions are caused by certain types of Robertsonian translocations. This is to be expected in carriers of 14q14q and 15q15q translocations (Žižka et al, 1977) and in carriers of 22q22q translocations (cf. Farah et al, 1975; Mameli et al, 1978). Although some 22 trisomic children have been born alive, none of them so far had a 22q22q translocation. The four pregnancies of a woman with 15q22q also ended in abortions (Fried et al, 1974).

The reproductive histories of families with a 13q14q translocation have as a rule been perfectly normal. However, some exceptions have been described, although it is often difficult to judge whether the reproductive troubles reported were in fact caused by the translocations or were incidental (cf. von Koskull and Aula, 1974).

Males with a DqDq translocation, usually between 13 and 14, have repeatedly been found to suffer from decreased fertility, often caused by oligospermia, although this is by no means a general feature in such carriers. Thus among 233 males with oligospermia, 8 DqDq translocation carriers were found, which amounts to 3.4 percent as compared with 0.1 percent in newborns (cf. Nielsen and Rasmussen, 1976). In their review of chromosomal causes of infertility of human males, Zuffardi and Tiepolo (1982) report that the incidence of Robertsonian translocations is ten times higher in such males than in the newborn.

The reason for the occasional infertility of DqDq translocation carrier males in not known. However, one wonders whether the translocation chromosomes in them might not be predominantly monocentric. In such chromosomes part of one long arm might be missing, possibly the cause of infertility, whereas in dicentric translocation chromosomes only heterochromatic short-arm material (and possibly the satellite stalks) are absent, or the number of transcribed ribosomal RNA genes may be too low, thus perhaps affecting the phenotype.

Nonrandomness of Robertsonian Translocations

Several hypotheses have been put forward to explain the relatively high incidence of Robertsonian translocations and the highly nonrandom participation of the different acrocentrics in them (cf. Hecht and Kimberling, 1971; Mikkelsen, 1973). One assumption is that centric heterochromatin and the satellite stalks are especially prone to breakage. Another phenomenon characteristic of acrocentric chromosomes is their participation in satellite associations, which may play an important role in bringing their centric regions together. It is of interest in this connection that in mouse cell lines the nucleolar chromosomes are preferentially involved in Robertsonian translocations (Miller et al, 1978).

Robertsonian translocations display features that distinguish them from other reciprocal translocations: (1) Ionizing radiation and chromosome-breaking substances do not seem to increase the incidence of Robertsonian translocations. On the other hand, certain substances, such as mitomycin C, preferentially cause these whole-arm transfers (Hsu et al, 1978). (2) Unlike other translocations, which take place more or less randomly between different chromosomes, the participation of the acrocentrics in Robertsonian translocations is extremely nonrandom (Table XXIV.1).

Most of the differences between Robertsonian and other translocations can be explained on the assumption that the former, as a rule, are the result of an exchange in meiosis or mitosis, and not of breakage and rejoining. Possibly the repeated sequences in the centric regions have an indiscriminate tendency to pair, and crossing-over in a reversely paired segment or a U-type exchange would lead to whole-arm transfers. This would be the mechanism producing the baseline number of Robertsonian translocations in the different classes (Table XXIV.1). The behavior of chromosomes 13, 14, and 21, on the other hand, is understandable if we assume that they have in common a homologous segment (A-B in Fig. XXIV.2), which is inverted in chromosome 14 relative to the two others. Crossing-over in the inverted region between chromosomes 13 and 14 and chromosomes 14 and 21, respectively, would give rise to t(13q14q) and t(14q21q), which are the common types of Robertsonian translocations.

The differences between the other classes can be understood on the basis of ascertainment bias. Chromosomes t(13q13q) and t(21q21q) come to our attention because the balanced carriers have only abnormal children, and the unbalanced ones are themselves abnormal. An increased number of Down's syndrome children also promote the detection of carriers of 21q22q, 13q21q and 15q21q translocations, whereas the relative rarity of translocations involving chromosome 22 (except 21q22q) obvi-

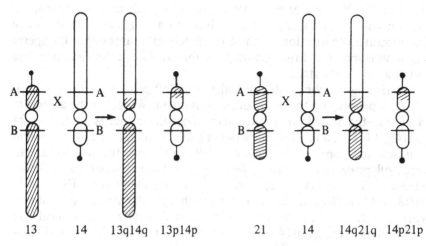

| 13 | 14 | 13q14q | 13p14p | | 21 | 14 | 14q21q | 14p21p |

Figure XXIV.2. Origin of Robertsonian translocations 13q14q and 14q21q through crossing-over in segment A-B that is inverted in chromosome 14 relative to chromosomes 13 and 21.

ously depends on the carrier's not producing abnormal children. Translocations between two chromosomes 14 were earlier thought to be non-existent (cf. Therman, 1980). However, during recent years four cases have come to light (Gracias-Espinal et al, 1982; Zhou et al, 1983; IA Uchida, unpublished; and probably Caspersson et al, 1971).

U-type exchange, instead of crossing-over, is the most plausible explanation for the formation of reverse tandem associations between two chromosomes 21 with one centromere inactivated and satellites on both ends (for example, Bartsch-Sandhoff and Schade, 1973; Schuh et al, 1974).

Segregation in Carriers of Robertsonian Translocations

The existing information on segregation ratios in DqDq and DqGq carriers is summarized in Table XXIV.2. All the offspring of a 13q13q and 21q21q carrier are 13 trisomic and 21 trisomic, respectively. Even a family with eight children with Down's syndrome in which the mother was a 21q21q carrier has been described (Furbetta et al, 1973).

It should be stressed that a carrier of a translocation between two homologous chromosomes can have only abnormal offspring or spontaneous abortions. Thus in a woman who was ascertained because she had four spontaneous abortions, the two chromosomes 7 had formed a t(7p7p) and a t(7q7q) (Niikawa and Ishikawa, 1983). Furthermore practically all carriers of t(14q14q), t(15q15q), and t(22q22q) have been identified because they have had only spontaneous abortions.

However, an unexpected outcome was observed in two families (Kirkels et al, 1980; Palmer et al, 1980). In both, a phenotypically normal woman with 45,XX,t(22q22q) gave birth to a daughter with the same chromosome constitution. In these extremely rare cases either the sperm was monosomic for chromosome 22 or the extra 22 in the daughters was lost at a very early stage.

The risk figures for producing unbalanced offspring are somewhat different, depending on the mode of ascertainment. Whatever the ascertainment bias, the probability of producing a healthy carrier is 50 percent for 13q14q, 14q21q, and 21q22q carriers (Table XXIV.2). In families found in unselected populations, such as newborn infants, the risk for unbalanced offspring is less than in families ascertained through an affected member. There is general agreement with the statement by Evans et al (1978, p. 112): "In conclusion, this and other similar studies suggest that when ascertained in a family by chance, both balanced reciprocal and Robertsonian, except t(14q21q), translocations carry a low risk of producing a congenitally malformed offspring."

For any carrier of a 13q14q translocation, the risk is very low. Hamerton (1970) estimated the risk for a 13qDq carrier to be 0.67 percent,

Table XXIV.2 Segregation in Carriers of Robertsonian Translocations.[a]

| Translocation | Risk of Unbalanced Offspring (%) | | Probability of Healthy Carrier (%) |
	Female carrier	Male carrier	
13qDq (mostly 14)	<1	Very low	50
13q13q	100	100	0
Dq21q (mostly 14)	10	Very low, estimated 2.4	50
21q22q	6.8	Upper limit 2.9	50
21q21q	100	100	0

[a]Data from Hamerton, 1970; Mikkelsen, 1971; and Chapman et al., 1973.

which is usually quoted as less than 1 percent. Low as this figure is, it is still higher than the risk for the general population, in which the estimates of the incidence of 13 trisomic children range from 0.005 percent to 0.02 percent. It should also be remembered that this risk figure (<1 percent) is based on one 13 trisomic offspring, who was not a proband (Hamerton, 1970).

Most information on segregation of Robertsonian translocations comes from families ascertained through an individual with Down's syndrome (Table XXIV.2). The risk figure for a female Dq21q carrier is about 10 percent. The risk for a male carrier is much lower. It has been estimated by Hamerton (1971) to be 2.4 percent, but this figure is based on a very small number of cases. The latest risk figure, 6.8 percent, for a female carrier of a t(21q22q) has been determined by Chapman et al (1973). For a male carrier the estimated upper limit for the risk is 2.9 percent.

The considerable difference in the risk of 13qDq and Dq21q carriers producing unbalanced offspring is assumed to depend on an alternate meiotic configuration being symmetrical for the former and asymmetrical for the latter (Fig. XXIV.3). Segregation in the configuration formed by a DqDq translocation would, as a rule, result in one cell with the translocation chromosome and another with the two free chromosomes. In the meiotic configuration formed by Dq21q, the free 21 would be near the same pole as the translocation and would therefore sometimes undergo adjacent segregation. Probably also because 21 is smaller than a D chromosome, it would more often fail to form a chiasma and drift at random in the first meiotic anaphase. This may explain the observation that, although the alternate meiotic configuration formed by 21q22q is symmetrical, the risk for unbalanced offspring is much higher than for a DqDq carrier. Another factor that may influence the empirical risk figures is the possibility that more 13 trisomic than 21 trisomic zygotes end up as abortions.

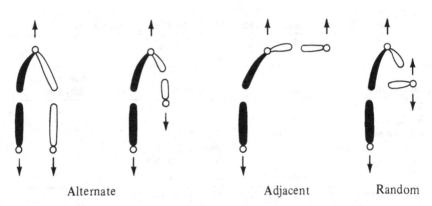

Alternate   Adjacent   Random

Figure XXIV.3. Different modes of meiotic orientation and segregation of Robertsonian translocations. Alternate segregation of t(DqDq) or t(DqGq) results in normals and carriers (1:1); adjacent segregation of t(DqGq) in trisomy and monosomy of G (1:1), and random drifting of univalent G in all four types (1:1:1:1).

An opportunity to determine directly the segregation also in carriers of Robertsonian translocations is provided by the technique of fertilizing hamster eggs with human sperm (Chapter IV). In an analysis of 24 sperm of a carrier of t(14q21q), 16 turned out to be chromosomally normal, 4 showed the balanced translocation; of the remaining 3, one had the translocation and was lacking chromosome 22, another had the translocation and an extra 21, whereas the third lacked a chromosome 21 (Balkan and Martin, 1983).

Interchromosomal Effects

By *interchromosomal effects* are meant the influence certain abnormal chromosomes may have on the behavior of nonhomologous chromosomes. Such effects have been established, for instance, in *Drosophila* (cf. Oksala, 1958).

In the human, interchromosomal effects have been claimed for various structurally abnormal chromosomes. Thus very large heterochromatic blocks, for instance in chromosome 9, have been thought to promote nondisjunction of other chromosomes (cf. Verma and Dosik, 1980). A similar claim has been made for inversions (cf. Kaiser, 1980). Reciprocal and Robertsonian translocations have also been reported to increase the risk of nondisjunction of other chromosomes. Thus, according to Mikkelsen (1971) a carrier of a t(DqDq) has a 2 percent risk of having a 21-trisomic child. On the other hand, Harris et al (1979), who analyzed segregation in 86 families with t(13q14q), found no chromosomally unbalanced offspring. Unfortunately the evidence for interchromosomal effects in the human is largely anecdotal, and this interesting question, which also has important implications for genetic counseling, should be studied more systematically than has been done so far.

References

Abe T, Morita M, Kawai K, et al (1975) Transmission of a t(13q22q) chromosome observed in three generations with segregation of the translocation D_1-trisomy syndrome. Humangenetik 30:207–215

Atkins L, Bartsocas CS (1974) Down's syndrome associated with two Robertsonian translocations, 45,XX, − 15, − 21, + t(15q21q) and 46,XX, − 21, + t(21q21q). J Med Genet 11:306–309

Balkan W, Martin RH (1983) Segregation of chromosomes into the spermatozoa of a man heterozygous for a 14;21 Robertsonian translocation. Am J Med Genet 16:169–172

Bartsch-Sandhoff M, Schade H (1973) Zwei subterminale Heterochromatin regionen bei einer seltenen Form einer 21/21-Translokation. Humangenetik 18:329–336

Caspersson T, Hultén M, Lindsten J, et al (1971) Identification of different Robertsonian translocations in man by quinacrine mustard fluorescence analysis. Hereditas 67:213–220

Chapman CJ, Gardner RJM, Veale AMO (1973) Segregation analysis of a large t(21q22q) family. J Med Genet 10:362–366

Cohen MM, Takagi N, Harrod EK (1968) Trisomy D_1 with two D/D translocation chromosomes. Am J Dis Child 115:185–190

Daniel A, Lam-Po-Tang PRLC (1976) Structure and inheritance of some heterozygous Robertsonian translocations in man. J Med Genet 13:381- 388

Evans JA, Canning N, Hunter AGW, et al (1978) A cytogenetic survey of 14,069 newborn infants. III. An analysis of the significance and cytologic behavior of the Robertsonian and reciprocal translocations. Cytogenet Cell Genet 20:96–123

Farah LMS, de S Nazareth HR, Dolnikoff M, et al (1975) Balanced homologous translocation t(22q22q) in a phenotypically normal woman with repeated spontaneous abortions. Humangenetik 28:357–360

Ford CE (1970) The population cytogenetics of other mammalian species. In: Jacobs PA, Price WH, Law P (eds) Human population cytogenetics. Williams and Wilkins, Baltimore, pp 221–239

Fried K, Bukovsky J, Rosenblatt M, et al (1974) Familial translocation 15/22. A possible cause for abortions in female carriers. J Med Genet 11:280–282

Furbetta M, Falorni A, Antignani P, et al (1973) Sibship (21q21q) translocation Down's syndrome with maternal transmission. J Med Genet 10:371–375

Gracias-Espinal R, Roberts SH, Duckett DP, et al (1982) Recurrent spontaneous abortions due to a homologous Robertsonian translocation (14q14q). J Med Genet 19:465–467

Hamerton JL (1970) Robertsonian translocations. In: Jacobs PA, Price WH, Law P (eds) Human population cytogenetics. Williams and Wilkins, Baltimore, pp 63–80

Hamerton JL (1971) Human cytogenetics I. Academic, New York

Harris DJ, Hankins L, Begleiter ML (1979) Reproductive risk of t(13q14q) carriers: case report and review. Am J Med Genet 3:175–181

Hecht F, Kimberling WJ (1971) Patterns of D chromosome involvement in human (DqDq) and (DqGq) Robertsonian rearrangements. Am J Hum Genet 23:361- 367

Hongell K, Gripenberg U, Iivanainen M (1972) Down's syndrome. Incidence of translocations in Finland. Hum Hered 22:7-14

Hook EB, Hamerton JL (1977) The frequency of chromosome abnormalities detected in consecutive newborn studies—differences between studies— results by sex and by severity of phenotypic involvement. In: Hook EB, Porter IH (eds) Population cytogenetics. Academic, New York, pp 63–79

Hsu TC, Pathak S, Basen BM, et al (1978) Induced Robertsonian fusions and tandem translocations in mammalian cell cultures. Cytogenet Cell Genet 21:86–98

Jacobs PA (1977) Structural rearrangements of the chromosomes in man. In: Hook EB, Porter IH (eds) Population cytogenetics. Academic, New York, pp 81–97

John B, Freeman M (1975) Causes and consequences of Robertsonian exchange. Chromosoma 52:123–136

Kaiser P (1980) Pericentrische Inversionen menschlicher Chromosomen. Thieme, Stuttgart

Kirkels VGHJ, Hustinx TWJ, Scheres JMJC (1980) Habitual abortion and translocation (22q;22q): unexpected transmission from a mother to her phenotypically normal daughter. Clin Genet 18:456–461

Koskull H von, Aula P (1974) Inherited (13;14) translocation and reproduction. Humangenetik 24:85–91

Mameli M, Cardia S, Milia A, et al (1978) A further case of a 22;22 Robertsonian translocation associated with recurrent abortions. Hum Genet 41:359–361

Mikelsaar A-V, Schmid M, Krone W, et al (1977) Frequency of Ag-stained nucleolus organizer regions in the acrocentric chromosomes of man. Hum Genet 37:73–77

Mikkelsen M (1971) Down's syndrome. Current stage of cytogenetic research. Humangenetik 12:1–28

Mikkelsen M (1973) Non-random involvement of acrocentric chromosomes in human Robertsonian translocations. In: Wahrman J, Lewis KR (eds) Chromosomes today. Vol 4. Wiley, New York, pp 253–259

Mikkelsen M, Basli A, Poulsen H (1980) Nucleolus organizer regions in translocations involving acrocentric chromosomes. Cytogenet Cell Genet 26:14–21

Miller OJ, Miller DA, Tantravahi R, et al (1978) Nucleolus organizer activity and the origin of Robertsonian translocations. Cytogenet Cell Genet 20:40–50

Niebuhr E (1972) Dicentric and monocentric Robertsonian translocations in man. Humangenetik 16:217–226

Nielsen J, Rasmussen K (1975) Extra marker chromosome in newborn children. Hereditas 81:221–224

Nielsen J, Rasmussen K (1976) Autosomal reciprocal translocations and 13/14 translocations: a population study. Clin Genet 10:161–177

Niikawa N, Ishikawa M (1983) Whole-arm translocation between homologous chromosomes 7 in a woman with successive spontaneous abortions. Hum Genet 63:85–86

Oksala T (1958) Chromosome pairing, crossing over, and segregation in meiosis in Drosophila melanogaster females. Cold Spring Harbor Symp Quant Biol 23:197–210

Orye E, Delire C (1967) Familial D/D and D/G$_1$ translocation. Helv Paediatr Acta 22:36–40

Palmer CG, Conneally PM, Christian JC (1969) Translocations of D chromosomes in two families t(13q14q) and t(13q14q)+(13p14p). J Med Genet 6:166-173

Palmer CG, Schwartz S, Hodes ME (1980) Transmission of a balanced homologous t(22q;22q) translocation from mother to normal daughter. Clin Genet 17:418–422

Pérez-Castillo A, Abrisqueta JA (1978) Patau's syndrome and 13q21q translocation. Hum Genet 42:327–331

Rockman-Greenberg C, Ray M, Evans JA, et al (1982) Homozygous Robertsonian translocations in a fetus with 44 chromosomes. Hum Genet 61:181–184

Schober AM, Fonatsch C (1978) Balanced reciprocal whole-arm translocation t(1:19) in three generations. Hum Genet 42:349–352

Schuh BE, Korf BR, Salwen MJ (1974) A 21/21 tandem translocation with satellites on both long and short arms. J Med Genet 11:297–299

Therman E (1980) Human chromosomes. Springer, New York

Verma RS, Dosik H (1980) Human chromosomal heteromorphisms: nature and clinical significance. Int Rev Cytol 62:361–383

Vianna-Morgante AM, Nunesmaia HG (1978) Dissociation as probable origin of mosaic 45,XX,t(15;21)/46,XY,i(21q). J Med Genet 15:305–310

Zhou H, Kang X, Zhang Q (1983) Homologous 14q14q Robertsonian translocation in man. Chin Med J 96:625–633

Žižka J, Balíćek P, Finková A (1977) Translocation D/D involving two homologous chromosomes of the pair 15. Hum Genet 36:123–125

Zuffardi O, Tiepolo L (1982) Frequencies and types of chromosome abnormalities associated with human male infertility. In: Crosignani PG, Rubin BL (eds) Genetic control of gamete production and function. Academic, London, pp 261–273

XXV

Double Minutes and Homogeneously Stained Regions

What are DMs and HSRs?

Double minutes (DM) and homogeneously stained regions (HSR) provide good examples of phenomena that were originally regarded as cytological oddities but that have turned out to be expressions of a fundamental process, gene amplification. Evidence for the structure and behavior of DMs and HSRs comes from two sources: human and animal cancers and cell lines subjected to selective pressure for drug resistance. So far they have not been observed in normal cells.

DMs are small spherical structures that occur in pairs, their number varying greatly from cell to cell. They were first described in the 1960s in solid human tumors and in induced mouse sarcomas (cf. Cowell, 1982). Many of the early findings were in children's neurogenic tumors, but gradually observations of their occurrence in most types of solid tumors and many leukemias have accumulated (cf. Sandberg, 1979).

HSRs are, as the name indicates, chromosome segments that are increased in length and stain uniformly with banding techniques. They were first described by Biedler and Spengler (1976) in human neuroblastoma cells and in methotrexate-resistant hamster cell lines. Since then, HSRs have been found in an increasing number of human and animal cancers. Both HSRs and DMs occur in addition to the usual chromosome complement of the cell.

Structure of Double Minutes

DMs vary from very small double dots (Fig. XXV.1a), which resemble diplococci, to larger spherical or fragment-like structures and rings (Fig. XXV.1b,c). Their staining properties show them to consist of chromosomal material, which, however, stains evenly with banding techniques.

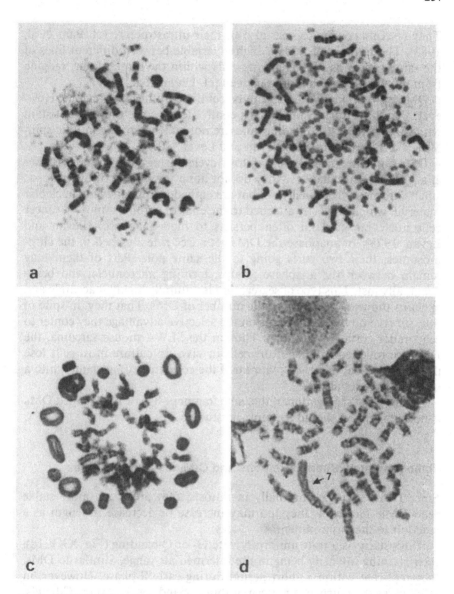

Figure XXV.1. G-banded metaphase plates from a human neuroblastoma cell line. (a) Small double minutes; (b) medium-sized double minutes; (c) large rings; (d) long, homogeneously stained region on chromosome 7 (Biedler JL, Ross RA, Shanskes, et al (1980) Human neuroblastoma cytogenetics: search for significance of homogeneously stained regions and double minute chromosomes. In: Evans AE (ed) Advances in neuroblastoma research. Raven, New York, pp 81–96).

They resemble chromosomes also in their ultrastructure (cf. Bahr et al, 1983). The number of DMs is highly variable between different lines of the same cancer as well as among cells within the same tumor, ranging from none to several hundred per cell (cf. Levan et al, 1981).

DMs were originally thought to be centric chromosome regions. However, they have turned out to lack centromeres, since they do not stain with C-banding or Cd-banding, and are not attached to the mitotic spindle (Barker and Hsu, 1978; Levan and Levan, 1978).

DMs replicate once in every mitotic cycle during early S phase (Barker et al, 1980a). In metaphase they do not lie in the middle of the plate, as small chromosomes usually do, but are embedded in persisting nucleolar material, which is mostly attached to the ends of chromosomes (in cancer cells nucleolar material often persists to meta-anaphase) (Levan and Levan, 1978). In anaphase the DMs get a free ride attached to the chromosomes, their two parts going to the same pole. Part of them may remain between the anaphase groups, forming micronuclei and being eventually lost. This type of segregation is naturally highly inaccurate and explains the wide variation in the number of DMs. That they, in spite of this, survive obviously depends on the selective advantage they confer to cells under certain conditions. Thus in the SEWA mouse sarcoma, the DMs are retained in 90% of the cells in vivo. In culture most cells lose the DMs, which, however, reappear if the cells are retransplanted into a mouse (Levan et al, 1977).

The mode of segregation of the large fragment-like or ring-shaped DMs is not known and should certainly be studied.

Homogeneously Stained Regions and C-Minus Chromosomes

Since HSRs segregate normally in mitosis, they are much more stable than DMs. However, they too may increase or decrease in length as a reaction to the environment.

HSRs stain, as a rule, uniformly with G- or Q-banding (Fig. XXV.1d), their staining intensity being in the intermediate range. Similar to DMs, they replicate within a short period during early S phase. However, in some cases they display a pattern of thin G-bands or they show C-bands, which obviously means that an additional segment has been coamplified with the repeated gene (cf. Cowell, 1982). Even a whole HSR in a human breast carcinoma cell line was intensively Q-bright, similar to Y-heterochromatin, and stained darkly with C-banding (Barker et al, 1980b). HSRs may be of sizable length, the DNA content of the chromosome involved being increased from 30 percent to 250 percent of the cell's DNA (Balaban-Malenbaum et al, 1979).

HSRs may involve different specific regions in the chromosome complement (Biedler and Spengler, 1976). Interestingly, several of the points in human neuroblastomas coincide with the hot spots for mitotic chiasmata in Bloom's syndrome (Kuhn et al, 1985). This probably means that these regions are especially liable to undergo unequal crossing-over.

HSRs involving the ribosomal RNA genes (satellite stalks) on chromosome 14 were observed in two families (cf. Cowell, 1982). Similar HSRs occur in three chromosomes in a rat hepatoma cell line (Miller et al, 1979). These structures consist of unstained satellite stalks alternating with typical HSRs. Instead of some 200 copies of ribosomal RNA genes present in normal rat cells, the hepatoma cells contain some 2,000 (Miller et al, 1979).

Levan et al (1981) found in some SEWA mouse sarcoma lines that the DMs were gradually being replaced by homogeneously stained telocentric or metacentric chromosomes, which varied in number from zero to fifteen per cell. These chromosomes did not show a C-band and were accordingly called C-minus or CM chromosomes. Obviously CMs are independent HSRs that have gained a centromere.

DMs and HSRs as Expressions of Gene Amplification

It is now clear that DMs, HSRs, and CMs are expressions of the same process: gene amplification. This was first proposed by Biedler and Spengler (1976), who observed that the HSRs appeared in hamster cells during an increase in their methotrexate resistance. This assumption has been amply confirmed.

The cancer drug methotrexate is a folic acid antagonist that is bound by the enzyme dihydrofolate reductase. Resistance to methotrexate can be induced in cultured mammalian cells by first exposing them to a low concentration of the drug, which kills most of them. The few survivors are grown out and exposed to a higher concentration. This process is repeated several times, the cells becoming stepwise more and more resistant to methotrexate (cf. Schimke, 1980). Correspondingly the number of dihydrofolate reductase genes increases. Simultaneously, the HSRs and DMs, which have been shown to consist of repeats of this gene, become longer or more numerous (cf. Schimke, 1980).

Cells resistant to several other substances, such as colchicine, vincristine, phosphonacetyl-L-aspartate (PALA), and cadmium, to mention a few, have also been observed to develop HSRs and DMs (cf. Cowell, 1982).

DMs and HSRs are Interchangeable

A variety of observations show that DMs and HSRs (or CMs) represent the same phenomenon in different disguises—DMs reflecting an unstable and HSRs a more stable state. First, as described above, they display similar staining and replication characteristics. Even more convincing is their behavior: they occur often in the same cancer, but not in the same cell. During the development of the SEWA mouse sarcomas, HSRs may vanish, to be replaced by DMs (or CMs). Or conversely, when DMs disappear, HSRs or CMs make their appearance (Levan et al, 1981). This has been demonstrated convincingly also in cell hybrids between a human neuroblastoma and mouse fibroblasts in which the HSRs on the two human chromosomes 1 were replaced by DMs (Balaban-Malenbaum and Gilbert, 1980).

Origin of DMs and HSRs

The origin of HSRs and DMs raises some intriguing questions. How does either type originally arise? And how do HSRs break up into DMs? An even more puzzling phenomenon is the integration of DMs into a chromosome to form an HSR. Neither is it known how the CMs pick up their C-bandless centromeres.

The mechanisms for the break-up of HSRs or integration of DMs are at present completely unknown, as is the origin of CM centromeres. Concerning the origin of the repeated sequences that make up these structures, some hypotheses have been put forward.

The most probable explanation for gene amplification is repeated unequal crossing-over, either between sister chromatids or homologous chromosomes (Chapter X) (cf. Therman and Kuhn, 1981). This clearly applies to the amplification of ribosomal RNA genes described above (Miller et al, 1979), since unequal crossing-over between satellite stalks obviously is not a too uncommon phenomenon (Therman et al, 1981). Whether unequal crossing-over alone suffices to account for the amazingly fast appearance—and disappearance—of HSRs is not clear.

An alternative mechanism that has been assumed to lead to gene amplification in microorganisms is saltatory replication (cf. Schimke, 1982). This means that a chromosome segment replicates more than once during an S period, and the additional copies are either integrated into the chromosome or become independent DMs.

Significance of HSRs and DMs

As shown above, DMs and HSRs (and CMs) are different expressions of gene amplification. The role played by them in the development of cells resistant to various drugs is obvious. The idea has also been proposed that this could be the mechanism inducing resistance in whole organisms to various agents, such as insecticide resistance in insects (cf. Schimke, 1980).

HSRs have been found to be expressions of oncogene amplification, which is a step in the malignant transformation of mammalian cells (Chapter XXVI). In already transformed cells, amplification of certain genes probably makes the cells more malignant and competitive (Chapter XXVII).

Since an HSR has to be relatively long to be microscopically visible, small HSRs may be much more common than has been assumed. It is possible that this type of gene amplification has played an important role in evolution by creating duplicate genes, gene clusters, and, combined with mutation, multigene families (cf. Schimke, 1980).

References

Bahr G, Gilbert F, Balaban G, et al (1983) Homogeneously staining regions and double minutes in a human cell line: chromatin organization and DNA content. JNCI 71:657–661

Balaban-Malenbaum G, Grove G, Gilbert F (1979) Increased DNA content of HSR-marker chromosomes of human neuroblastoma cells. Exp Cell Res 119:419-423

Balaban-Malenbaum G, Gilbert F (1980) The proposed origin of double minutes from homogeneously staining region (HSR)-marker chromosomes in human neuro-blastoma hybrid cell lines. Cancer Genet Cytogenet 2:339–348

Barker PE, Drwinga HL, Hittelman WN, et al (1980a) Double minutes replicate once during S phase of the cell cycle. Exp Cell Res 130:353–360

Barker PE, Hsu TC (1978) Are double minutes chromosomes? Exp Cell Res 113:457–458

Barker PE, Lau Y-F, Hsu TC (1980b) A heterochromatic homogeneously staining region (HSR) in the karyotype of a human breast carcinoma cell line. Cancer Genet Cytogenet 1:311–319

Biedler JL, Ross RA, Shanske S, et al (1980) Human neuroblastoma cytogenetics: search for significance of homogeneously stained regions and double minute chromosomes. In: Evans AE (ed) Advances in neuroblastoma research. Raven, New York, pp 81–96

Biedler JL, Spengler BA (1976) Metaphase chromosome anomaly: association with drug resistance and cell-specific products. Science 191:185–187

Cowell JK (1982) Double minutes and homogeneously staining regions: gene amplification in mammalian cells. Annu Rev Genet 16:21–59

Kuhn EM, Therman E, Denniston C (1985) Mitotic chiasmata, gene density, and oncogenes. Hum Genet 70:1–5

Levan A, Levan G (1978) Have double minutes functioning centromeres? Hereditas 88:81–92

Levan A, Levan G, Mandahl N (1981) Double minutes and C-bandless chromosomes in a mouse tumor. In: Arrighi FE, Rao PN, Stubblefield E (eds) Genes, chromosomes, and neoplasia. Raven, New York, pp 223–251

Levan G, Mandahl N, Bengtsson BO, et al (1977) Experimental elimination and recovery of double minute chromosomes in malignant cell populations. Hereditas 86:75–90

Miller OJ, Tantravahi R, Miller DA, et al (1979) Marked increase in ribosomal RNA gene multiplicity in a rat hepatoma cell line. Chromosoma 71:183–195

Sandberg AA (1979) The chromosomes in human cancer and leukemia. Elsevier North Holland, New York

Schimke RT (1980) Gene amplification and drug resistance. Sci Am 243:60–69

Schimke RT (1982) Summary. In: Schimke RT (ed) Gene amplification. Cold Spring Harbor Laboratory, New York, pp 317–333

Therman E, Kuhn EM (1981) Mitotic crossing-over and segregation in man. Hum Genet 59:93–100

Therman E, Otto PG, Shahidi NT (1981) Mitotic recombination and segregation of satellites in Bloom's syndrome. Chromosoma 82:627–636

XXVI
Chromosomes and Oncogenes

What is Cancer?

There are two types of neoplasms, both of which are expressions of abnormal growth. Benign tumors are outgrowths that are self-limiting, that is, they grow to a certain size and then stop or regress. Most of us are acquainted with benign tumors, such as polyps or warts. Malignant tumors, on the other hand, usually show unlimited growth; they escape the rules of differentiation and grow wild. They also have the ability to infiltrate and destroy normal tissues, and most malignant tumors are also capable of spreading to new sites by metastasis. Tissue cultures of cancer cells are immortal, whereas normal cells divide only a limited number of times in vitro. Furthermore, normal cultured cells avoid contact, whereas malignant cells are able to grow in several layers. The histological structure of cancerous tumors usually differs greatly from the original normal tissue and often appears to be relatively disorganized and anaplastic. Advanced cancer may display a varied cytological picture; cells with small and large, often giant, nuclei exist side by side with cells having several or weirdly shaped nuclei. Cancer mitoses also exhibit a wide spectrum of abnormalities, such as multiple poles, disorganized metaphases, anaphases with laggards and chromatid bridges, endomitoses, and endoreduplications. These phenomena indicate that the mitotic processes of chromosome replication and division, usually so orderly, have been dramatically disrupted.

Malignant tumors are by no means specific to man but are found throughout the animal kingdom, from ants to whales. Corresponding phenomena also occur in plants. The so-called crown gall, for example, which affects a wide variety of plant species, resembles animal cancer.

The induction of crown gall requires that a wound in a suitable host plant be inoculated with *Agrobacterium tumefaciens*. After enough cells have been transformed into crown gall cells, the tumor continues to grow even after the bacteria have been killed. The transformation occurs when a segment of a plasmid is transferred from the inducing bacterium and is incorporated into the host cell genome.

Cancer Induction

Cancer is caused by mutation in the broadest sense of the word. There is convincing evidence that the overwhelming majority of malignant tumors have a clonal origin (cf. Nowell, 1976), a change that takes place in a single cell from which the entire tumor is derived. In primary tumors all cells may display the same abnormal chromosome constitution. Enzyme studies of women with different alleles of an X-linked gene show that the same X chromosome is active in all the cells of a tumor. The immunoglobulin chain produced by a plasma-cell tumor nearly always confirms the assumption of a clonal origin of the disease.

That an organism's genetic constitution may have a decisive effect on its probability of developing cancer is borne out by many observations. For instance, there are inbred mouse strains in which most animals develop malignant tumors. A definite tendency to develop malignant disease seems to be inherited in some human families. In many such "cancer syndrome" families about one-half the members eventually develop one or more malignancies; this suggests that a dominant gene is the determining factor. In one family (Lynch et al, 1977), 20 of 88 relatives studied had been affected with carcinoma of the colon and/or endometrium (uterine epithelium). Other types of tumors were also found in the family, and 16 individuals had more than one primary malignancy. For offspring of affected parents the risk of colon/endometrial cancer was 52.8 percent in the 20- to 60-year age group, whereas none of the children of unaffected parents or the unrelated spouses developed cancer. It would be of great interest to know whether the frequency of chromosome aberrations is increased in the somatic cells of those bearing the "cancer gene."

Miller (1967) reviewed populations with especially high risks of leukemia. In Caucasian children under 15, the incidence of leukemia is 1/2880. If one identical twin has leukemia, the risk of leukemia for the other twin is 1/5. In persons who have been exposed to ionizing radiation, the risk is also greatly increased. Thus, in Hiroshima survivors who were within 1000 meters of the hypocenter, the probability of developing leukemia is 1/60.

Especially interesting are the chromosome breakage syndromes (Chapter IX), of which the most extensively studied are ataxia telangiectasia, Fanconi's anemia, and Bloom's syndrome. The homozygotes, which show a greatly increased rate of chromosome breakage, also have a greatly increased risk for malignant disease; for instance, in Bloom's syndrome it is 1/4 of patients, and in ataxia it is 1/8.

It is obvious that malignant disease depends to a great extent on genetic factors. It is also known that most, possibly all, carcinogens are mutagens; however, not all mutagens necessarily induce cancer. There is also a direct connection between the ability of various agents to break chromosomes and their carcinogenicity. Examples of such agents include ionizing radiation, a great variety of chemicals, chromosome breakage syndromes, and viruses.

Interestingly, the breakpoints in the constant chromosome aberrations in cancer seem in a significant number of cases to coincide with the fragile chromosome sites (cf. Yunis, 1983; Yunis and Soreng, 1984; Hecht and Sutherland, 1984).

The many facets of cancer research have come together in the recent discovery of oncogenes and their relation to chromosome abnormalities. The fascinating results obtained have become possible through close coordination of cytogenetic, viral, and molecular studies. The field of oncogenesis is undergoing such rapid development that present ideas are bound to undergo rapid and drastic changes. The following is, therefore, a tentative attempt to survey oncogene research as it stands now.

Oncogenes

So-called proto-oncogenes are normal genes present in all metazoan cells. What roles they play under normal circumstances is mostly unknown. It is likely—and for a couple of oncogenes there is evidence for this—that they are involved in cell division, growth, and differentiation (cf. Bishop, 1983). Some of the proto-oncogenes produce tyrosine kinases which phosphorylate proteins (cf. Land et al, 1983).

Viruses have for a long time been known to cause cancer in animals. The Rous sarcoma virus, a retrovirus (RNA virus) that induces malignant tumors in chickens, was discovered as early as 1911. An association between virus infection and several human malignant diseases has been assumed. Such diseases include various leukemias, Hodgkin's disease, Burkitt's lymphoma, and cervical, hepatocellular, and nasopharyngeal cancers. Possibly virus infection constitutes one step in the multistep process of malignant transformation in these diseases.

Genes homologous to cellular proto-oncogenes are found in retroviruses known to cause cancer in various animal species. They transform cells either by being inserted into the host genome or by being present in multiple copies in the host cell (cf. Bishop, 1983). It is thought that the retroviruses picked up the oncogenes from the metazoan cells they had infected. A normal cell is transformed when one or more proto-oncogenes in it are activated, which can occur through mechanisms such as point mutation, position effect, or amplification.

Table XXVI.1. Examples of Oncogenes, Their Location on a Human Chromosome, and the Corresponding Virus (cf. de la Chapelle and Berger, 1983; Yunis, 1983).

Oncogene	Human Chromosome Location	Virus and Species of Origin
Blym-1	1p32	Not in a virus. Isolated from chicken lymphomas.
N-ras	1	Not in a virus. From *human* tumors.
raf-1	3p25	3611—murine sarcoma virus (mouse)
raf-2P	4	MC 29 myelocytomatosis virus (chicken)
fms	5q34	McDonough feline sarcoma virus (cat)
HH-ras-1	6p23-q12	Kirsten murine sarcoma virus (rat)
myb	6q23	Avian myeloblastosis virus (chicken)
erb-B	7pter-q22	Avian erythroblastosis virus (chicken)
mos	8q22	Moloney murine sarcoma virus (mouse)
myc	8q24	MC 29 myelocytomatosis virus (chicken)
abl	9q34	Abelson murine leukemia (mouse)
HH-ras	11p14	Harvey murine sarcoma virus (rat)
HK-ras-2	12p12-q24	Kirsten murine sarcoma virus (rat)
fes	15q26	Snyder-Theilen feline sarcoma virus (cat)
erb-A	17p11-q21	Avian erythroblastosis virus (chicken)
src	20	Rous sarcoma virus (chicken)
sis	22q13	Simian sarcoma virus (woolly monkey)
H-ras-2	X	Harvey murine sarcoma virus (rat)

The oncogene of the Rous sarcoma virus is called v-scr, and its normal homolog in a metazoan cell c-scr. The versions of this gene in fishes, birds, and mammals are closely related to the viral gene.

Some 20 different human proto-oncogenes are now known, and several of them have been localized to a specific chromosome band, or at least to a specific chromosome (Table XXVI.1). This has been done through somatic cell hybridization analyses or through in situ hybridization of a radioactive probe of a cloned c-onc gene, or of a viral v-onc gene, to human chromosomes (cf. Rowley, 1983; Yunis, 1983).

These types of oncogenes are dominant in their effects, as shown by transfection experiments. Purified DNA fragments from a variety of cancers, when transferred to recipient cells (a process called transfection), transform nonneoplastic cells with high efficiency. DNA fragments from normal cells also accomplish transformation, although with a very low frequency, which has been interpreted to mean that the fragmentation sometimes activates an oncogene (cf. Cooper, 1982).

Reciprocal Translocations and Oncogenes

It has long been thought that if the same chromosomal aberration is found consistently in a certain type of malignant disease, it is involved in its origin (cf. Sandberg, 1983). The search for such chromosome aber-

Table XXVI.2. Neoplasms with a Known Specific Recurrent Chromosomal
Defect (Modified from Yunis, 1984).

Disease	Chromosome Defect	Breakpoints
Leukemias		
Chronic granulocytic leukemia	t(9;22)	9q34.1 and 22q11.21
Acute nonlymphocytic leukemia		
M1	t(9;22)	9q34.1 and 22q11.21
M2	t(8;21)	8q22.1 and 21q22.3
M1, M2	t(6;9)	6p22p23 and 9q34
M1, M2	inv 3	3q21 and 3q26.2
M3	t(15;17)	15q22 and 17q11.2
M2, M4, M5b	inv16	16p13.2 and 16q22.1
M2, M4, M5a	t(9;11)	9p22 and 11q23
M1, M2, M4, M5, M6	del 5q	5q13 and 5q31
	del 7q	7q22 and 7q32
	+8	
Chronic lymphocytic leukemia	+12	
	t(11;14)	11q13 and 14q32
Acute lymphocytic leukemia		
L1-pre B cell	t(1;19)	1q23 and 19p13?
L1-T cell	t(11;14)	11p13-15 and 14q11-13
L1-L2	t(9;22)	9q34.1 and 22q11.21
L2	t(4;11)	4q21 and 11q23
L3	t(8;14)	8q24.13 and 14q32.33
Lymphomas		
Burkitt's, small noncleaved cell (non-Burkitt), large cell immunoblastic	t(8;14)	8q24.13 and 14q32.33
Follicular small cleaved, follicular mixed, and follicular large cell	t(14;18)	14q32.3 and 18q21.3
Small cell lymphocytic	+12	
	t(11;14)	11q13 and 14q32
Carcinomas		
Neuroblastoma, disseminated	del 1p	1p31p36
Small cell lung carcinoma	del 3p	3p14p23
Papillary cystadenocarcinoma of ovary	t(6;14)	6q21 and 14q24
Ewing sarcoma	t(11;22)	q24 and q12
Adenocarcinoma of the colon	del 12	q22q24
Constitutional retinoblastoma	del 13q	13q14.13
Retinoblastoma	del 13q	13q14
Aniridia-Wilms' tumor	del 11p	11p13
Wilms' tumor	del 11p	11p13
Benign solid tumors		
Mixed parotid gland tumor	t(3;8)	3p25 and 8q21
Meningioma	−22	

rations has been made difficult by the tendency of malignant cells to accumulate chromosome abnormalities *after* transformation has taken place (Chapter XXVII). Real progress in this field began only after 1970, when banding techniques were developed, and especially after high-resolution banding came into use (cf. Yunis, 1981). Other important technical improvements include short-term culture and synchronization of cancer cells (cf. Yunis, 1981, 1984).

The list of chromosome aberrations specific for specific types of malignant disease is continuously growing. In leukemias and lymphomas these are mostly balanced reciprocal translocations; in solid tumors they are deletions and sometimes trisomy (Table XXVI.2) (cf. de la Chapelle and Berger, 1983; Rowley, 1983; Yunis, 1983, 1984).

The first chromosome aberration consistently associated with malignancy, the Philadelphia chromosome (Ph[1]) (a G chromosome with about half of its long arm missing), was described in chronic myelogenous leu-

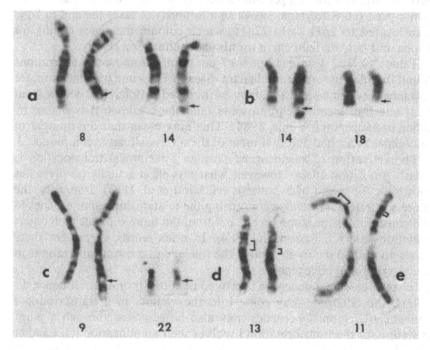

Figure XXVI.1. Selected Giemsa-banded chromosomes prepared by high-resolution technique at the 850 and 1200 band stages from patients with (a) non-Burkitt small cell lymphoma and t(8;14); (b) follicular small cleaved cell lymphoma and t(14;18); (c) chronic myelogenous leukemia and t(9;22); (d) constitutional retinoblastoma and partial loss of band 13p14; and (e) constitutional Wilms' tumor and partial loss of band 11p13. Arrows indicate breakpoint involved in the translocations, and brackets illustrate band deletion (Yunis, JJ. The chromosomal basis of human neoplasia. Science 221:227–236. Copyright 1983 by the AAAS).

kemia (CML) by Nowell and Hungerford (1960). Banding showed this chromosome to be 22, and a careful analysis by Rowley (1973) revealed that the abnormality in 90% of the cases was a translocation between 9q and 22q. With prophase banding the breakpoints were defined as q34.1 and q11.2 respectively (Fig. XXVI.1) (cf. Yunis, 1983).

With recombinant DNA techniques the oncogene c-abl (Table XXVI.1) has been localized to the part of 9q that is translocated to 22, in which it has come to lie next to the immunoglobulin light chain lambda gene; the significance of this is unknown.

Even more intensively studied has been the c-myc oncogene, which is possibly involved in the origin of the highly malignant Burkitt's lymphoma. The most common chromosome finding in malignant cells of patients with this disease is t(8;14)(q24;q32.3) (Figs. XXVI.1, XXVI.2). The c-myc has been mapped to 8q24, and the immunoglobulin heavy chain locus to 14q32.3. The new location of the c-myc gene next to the broken immunoglobulin gene in some way changes the regulation of the c-myc gene (cf. Robertson, 1984). In a minority of cases the end of 8q is translocated to 2p11 or to 22q11, which contain the genes coding for kappa and lambda light chain immunoglobulins respectively.

Table XXVI.2 lists examples of constant chromosome aberrations found in different types of malignant disease. The same translocation, for instance between 8q and 14q, may be involved in different diseases, or at least one translocation chromosome (and the breakpoint) is shared by different cancers (cf. Yunis, 1984). This may mean that the number of oncogenes is limited and that most of them have already been found.

The activation of an oncogene through a reciprocal translocation is clearly a position effect. However, what this effect actually involves lies mostly in the realm of hypothesis (cf. Land et al, 1983). Ironically, the more accurate observations concerning the relationships of oncogenes to the constant chromosome breaks become, the more difficult their interpretations are (cf. Robertson, 1984). In other words, at present there exists no unified theory to explain the role of chromosome aberrations in the activation of oncogenes.

In some cases the oncogene is moved next to a promoter sequence. In others, the oncogene may come into the vicinity of a transcriptional "enhancer" region. Oncogenes may also be activated through a point mutation, or the transformation involves their amplification (cf. Land et al, 1983).

In many primary tumors, as well as cultured malignant cells, homogeneously stained chromosome regions and double minutes have been observed (Chapter XXV). Usually the amplification happens during tumor development and is not involved in its origin. However, in colon carcinoma the HSRs and DMs reflect the amplification of the c-myc gene (Alitalo et al, 1983). In this cancer and in a few others, gene amplification

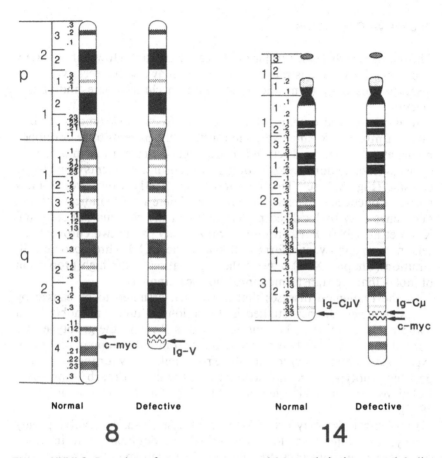

Figure XXVI.2. Location of c-myc oncogene and heavy chain immunoglobulin variable (V) and constant u (Cμ) genes on normal and defective chromosomes 8 and 14 in Burkitt's lymphoma, represented at the 1200 Giemsa band stage. The defective chromosome 8 loses the c-myc and gains V genes. The defective chromosome 14 gains c-myc from chromosome 8, becoming contiguous or near to Cμ. Arrows point to the normal and rearranged locations of these genes. Broken ends of defective chromosomes indicate breakpoint sites (Yunis, JJ. The chromosomal basis of human neoplasia. Science 221:227–236. Copyright 1983 by the AAAS).

may be the mechanism of transformation (cf. Marx, 1984) or at least constitute one step in carcinogenesis. Trisomy for a chromosome or chromosome segment may naturally have the same effect (cf. Gilbert, 1983).

Interestingly, constant chromosome aberrations have been found also in some types of benign tumors. Thus, in meningiomas, chromosome 22 is often missing (Zankl and Zang, 1980), and various translocations have been observed in mixed salivary gland tumors (Mark et al, 1983).

Recessive Oncogenes

The oncogenes discussed so far have been dominant. However, a variety of observations demonstrate that oncogenes may also be recessive. In the diseases caused by them, the most common chromosome finding is a deletion.

In Bloom's syndrome, in which the most frequent chromosome aberration is mitotic crossing-over, one-fourth of the patients have developed malignant disease. As discussed in Chapter X, segregation in a mitotic chiasma leads to homozygosity of the chromosome segments distal to the chiasma (Fig. X.3) (cf. Therman and Kuhn, 1981), which would allow recessive oncogenes to express themselves. There are also indications that oncogenes may be localized preferentially at the chiasma hot spots (cf. Kuhn et al, 1985). Furthermore, homozygosity of the two chromosome sets in complete hydatidiform moles with the 46,XX chromosome constitution (Chapter XXI) may be the explanation of the high proportion of moles that are transformed into choriocarcinomas.

Of the malignant diseases that have been assumed to be caused by recessive genes, the best studied is the childhood tumor retinoblastoma (cancer of the retina). The gene involved is Rb, and the genotype of a normal person is Rb/Rb. A prerequisite for tumor development is homozygosity or hemizygosity for the allele rb or nullosomy for the locus. The possible genotypes of retinoblastoma cells are thus rb/rb, rb/− or −/−. Deletions have placed the gene at 13q14 (Fig. XXVI.1) (cf. Cavenee et al, 1983).

Retinoblastoma may occur as isolated cases, usually involving only one eye; or the tendency to it is inherited as a dominant trait, in which case, as a rule, both eyes are affected. The genotype of a person with inherited retinoblastoma is either Rb/rb or Rb/−, and mutation or deletion of the Rb gene results in malignant transformation of a retinal cell. In persons with the genotype Rb/rb or Rb/−, the risk of developing retinoblastoma is some 100,000 times higher than in the population in general, and the risk for other types of cancer is also increased (cf. Murphree and Benedict, 1984).

Several mechanisms leading to retinoblastoma in persons with the genotypes Rb/rb or Rb/− have been established. Chromosome 13 may be lost, or a segment including 13q14 be deleted, or the chromosome with Rb be lost and the homolog duplicated. Other possibilities include mitotic recombination or the loss of the Rb allele through a translocation break (cf. Cavenee et al, 1983).

In a few persons who have developed retinoblastoma, all cells have a congenital heterozygous deletion of 13q14. These individuals are slightly retarded and have other physical anomalies (cf. Francke and Kung, 1976).

Observations similar to those in retinoblastoma have been made in other solid tumors. Thus, in patients with Wilms' tumor the band 11p13 is often missing (Fig. XXVI.1), showing the location of the gene responsible for this disease (cf. Solomon, 1984). In neuroblastoma the segment 1p31-p36 is deleted.

Multistep Carcinogenesis

The idea that malignant transformation occurs in several steps is not new (cf. Knudson, 1973; Fialkow et al, 1981). This hypothesis has been based on several lines of evidence. Malignant disease often appears 10 to 30 years after a person has been exposed to a carcinogen. Precancerous conditions, such as papillomas, adenomas, and preleukemic disorders, often precede transformation. Substances which alone do not induce cancer but promote the effect of carcinogens are known. Furthermore, it is now obvious that the activation of one oncogene as a rule is not enough to transform a cell, but is only one of several necessary steps (cf. Gilbert, 1983). Thus Burkitt's lymphoma may require three or four steps, including the activation of two oncogenes and infection with Epstein-Barr virus (cf. Land et al, 1983). Also, contrary to normal cells, which are able to divide a limited number of times, in cell lines with unlimited growth only one step is needed for transformation. This means that such cell lines may already have undergone the other necessary changes (cf. Cooper, 1982).

References

Alitalo K, Schwab M, Lin CC, et al (1983) Homogeneously staining chromosomal regions contain amplified copies of an abundantly expressed cellular oncogene (c-*myc*) in malignant neuroendocrine cells from a human colon carcinoma. Proc Natl Acad Sci USA 80:1707–1711
Bishop JM (1983) Cellular oncogenes and retroviruses. Annu Rev Biochem 52:301–354
Cavenee WK, Dryja TP, Phillips RA, et al (1983) Expression of recessive alleles by chromosomal mechanisms in retinoblastoma. Nature 305:779-784
Chapelle A de la, Berger R (1984) Report of the committee on chromosome rearrangements in neoplasia and on fragile sites. Cytogenet Cell Genet 37:274–311
Cooper GM (1982) Cellular transforming genes. Science 217:801–806
Fialkow PJ, Martin PJ, Najfeld V, et al (1981) Evidence for a multistep pathogenesis of chronic myelogenous leukemia. Blood 58:158–163
Francke U, Kung F (1976) Sporadic bilateral retinoblastoma and 13q− chromosomal deletion. Med Pediat Oncol 2:379–385
Gilbert F (1983) Chromosomes, genes, and cancer: a classification of chromosome abnormalities in cancer. J Nat Cancer Inst 71:1107–1114
Hecht F, Sutherland GR (1984) Fragile sites and cancer breakpoints. Cancer Genet Cytogenet 12:179–181
Knudson AG (1973) Mutation and human cancer. Adv Cancer Res 17:317–352
Kuhn EM, Therman E, Denniston C (1985) Mitotic chiasmata, gene density, and oncogenes. Hum Genet 70:1-5

Land H, Parada LF, Weinberg RA (1983) Cellular oncogenes and multistep carcinogenesis. Science 222:771–778

Lynch HT, Harris RE, Organ CH Jr, et al (1977) The surgeon, genetics, and cancer control: the cancer family syndrome. Ann Surg 185:435–440

Mark J, Dahlenfors R, Ekedahl C (1983) Cytogenetics of the human benign mixed salivary gland tumour. Hereditas 99:115–129

Marx JL (1984) Oncogenes amplified in cancer cells. Science 223:40–41

Miller RW (1967) Persons with exceptionally high risk of leukemia. Cancer Res 27:2420–2423

Murphree AL, Benedict WF (1984) Retinoblastoma: Clues to human oncogenesis. Science 223:1028–1033

Nowell PC (1976) The clonal evolution of tumor cell populations. Science 194:23–28

Nowell PC, Hungerford DA (1960) A minute chromosome in human chronic granulocytic leukemia. Science 132:1497

Robertson M (1984) Message of myc in context. Nature 309:585–587

Rowley JD (1973) A new consistent chromosomal abnormality in chronic myelogenous leukemia identified by quinacrine fluorescence and Giemsa staining. Nature (Lond) 243:290–293

Rowley JD (1983) Human oncogene locations and chromosome aberrations. Nature 301:290–291

Sandberg AA (1983) A chromosomal hypothesis of oncogenesis. Cancer Genet Cytogenet 8:277–285

Solomon E (1984) Recessive mutation in aetiology of Wilms' tumour. Nature 309:111–112

Therman E, Kuhn EM (1981) Mitotic crossing-over and segregation in man. Hum Genet 59:93–100

Yunis JJ (1981) New chromosome techniques in the study of human neoplasia. Hum Pathol 12:540–549

Yunis JJ (1983) The chromosomal basis of human neoplasia. Science 221: 227–236

Yunis JJ (1984) Clinical significance of high resolution chromosomes in the study of acute leukemias and non-Hodgkins lymphomas. In: Fairbanks VF (ed) Current hematology, Vol 3. Wiley, New York, pp 353–391

Yunis JJ, Soreng AL (1984) Constitutive fragile sites and cancer. Science 226:1199–1204

Zankl H, Zang KD (1980) Correlations between clinical and cytogenetical data in 180 human meningiomas. Cancer Genet Cytogenet 1:351–356

XXVII
Chromosomal Development of Cancer

Chromosomes and Cancer Progression

It has been known since the beginning of the present century that the number of chromosomes in cancer cells often deviates greatly from the usual number in healthy cells of the host organism. The prominent German biologist Theodor Boveri observed that multipolar mitoses in sea urchin eggs led to abnormal chromosome numbers, and these in turn led to abnormal development of the larvae. Since cancer cells often display multipolar divisions, Boveri concluded that the resulting deviant chromosome numbers were the cause of cancerous growth. Boveri's book *Zur Frage der Entstehung maligner Tumoren* ("On the problem of the origin of malignant tumors") appeared in 1914 (cf. Wolf, 1974); however, as has so often happened in cancer research, Boveri had put the cart before the horse. It is now clear that multipolar divisions appear only *after* the cells have undergone a malignant transformation. But Boveri's basic hypothesis, that chromosome aberrations may cause cancer, is very much alive today, as shown by the discovery of the relationship of chromosome structural changes to oncogenes (Chapter XXVI). It is now clear that, although certain constant chromosome anomalies are involved in the origin of cancer, most of the observed aberrations, both numerical and structural, arise during the progression of a malignant disease, and they in turn affect the further development of the tumor or leukemia. The difficulty has often been to distinguish the primary chromosome change from the secondary ones, and even now it is often unclear whether a certain abnormality is a step in carcinogenesis or only makes the already transformed cells more malignant.

Chromosome Studies in Ascites Tumors

Cancer cytology was launched in the 1950s with the study of transplantable ascites tumors of the mouse and the rat (cf. Yosida, 1975). These tumors were also the first mammalian tissues from which satisfactory chromosome preparations were obtained. Bayreuther (1952), who seems to have been the first investigator to apply a colchicine derivative to two mouse ascites tumors, observed that both the chromosome number and morphology deviated from those of normal mouse cells.

Subsequent studies have shown that many mouse ascites tumors are near-triploid or near-tetraploid and that the karyotype shows many morphologically abnormal chromosomes. Thus the hypotetraploid Ehrlich tumor, so widely used in various experiments, displays only a couple of chromosomes that can be matched, even approximately, with any normal mouse chromosome.

Primary tumors in man sometimes induce the development of ascites fluid, in which dividing tumor cells can be studied. Such cells also show striking abnormalities, both in chromosome number and structure.

Chromosome Studies in Primary Tumors

A vast number of primary tumors and leukemic conditions have been analyzed cytologically. The results have been reviewed in numerous articles and books (for example, Atkin, 1974; Sandberg and Sakurai, 1974; Makino, 1975; Sandberg, 1979; Nowell, 1982). The majority of malignant tumors have chromosome complements that are abnormal, both in chromosome number and structure. Many tumors display *marker chromosomes* that are morphologically abnormal (cf. Levan et al, 1977). Various strains of so-called HeLa cells, which originated from a human cervical cancer, are widely grown in tissue culture. The HeLa lines are near-triploid and exhibit many structurally changed chromosomes (cf. Heneen, 1976).

Most of the cells in a tumor belong to one *stemline*, which consists of cells with the same chromosome constitution, often including striking marker chromosomes. One or more cell lines with other chromosome constitutions may also be present. A stemline often responds to a new environment, for instance after transplantation into an alien host species, by a change in its chromosome constitution. (Some tumors have lost their immunological specificity to the extent that they are even able to grow in a different host species.) Chemotherapy may also affect the karyotype of a stemline (cf. Yosida, 1975). Finally, the stemline usually changes during the progression of a tumor. As an interesting oddity it may be mentioned that, in several types of leukemia, stemlines or individual cells with near-haploid chromosome constitutions never encountered elsewhere have been found (cf. Sandberg et al, 1982). The wide variation in chromosome constitution within a tumor and between neoplasias is reflected in the variable DNA content of nuclei, which has been determined by both spectrophotometry and flow cytometry.

The participation of the different chromosomes in the abnormalities of human cancer stemlines is nonrandom (cf. Rowley, 1977; Mitelman and Levan, 1978, 1981). Of the 24 different human chromosomes only the following 15 participated significantly in the anomalies in 1,871 cases of malignant disease: 1, 3, 5, 6, 7, 8, 9, 11, 12, 13, 14, 17, 20, 21, and 22 (Mitelman and Levan, 1981).

The chromosome evolution of individual malignancies is also nonrandom. Thus, in chronic myelogenous leukemia (CML), in which the primary aberration is t(9q;22q), other anomalies appear during the progression of the disease.

Mitelman et al (1976) (Fig. XXVII.1) listed additional chromosome changes in 66 of 200 CML patients who had the Ph[1] chromosome. In 88 percent of the cases the chromosomal development took what the authors called the major route. In 18 patients, a second Ph[1] chromosome appeared, in seven an extra chromosome 8, and in nine patients an i(17q) was found. In nine patients both two Ph[1] chromosomes and an extra 8

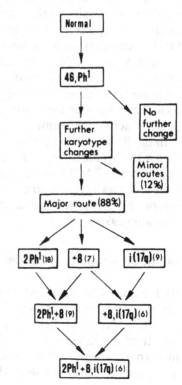

Figure XXVII.1. Chromosome abnormalities in addition to the Ph[1] chromosome in 66 cases of chronic myeloid leukemia. Figures in parentheses in the lower part of the diagram indicate the number of cases showing the chromosome abnormality (Mitelman et al, 1976).

occurred, whereas in six others the extra 8 was combined with an i(17q). Finally, in six patients the cells showed, in addition to two Ph[1] chromosomes, both an extra 8 and i(17q). The same type of development is often found when the disease is followed in individual patients (cf. de Grouchy and Turleau, 1974).

As discussed in Chapter XXV, amplification of certain chromosome segments may take place during tumor development, marked by the appearance of homogeneously stained chromosome regions and double minutes.

Chromosome analysis of tumors, especially of leukemias and lymphomas, may be a useful adjunct for diagnosis; it also provides clues for prognosis of the disease and its reaction to different therapeutic treatments (cf. Yunis, 1984).

The Apparent Predetermination of Chromosome Changes in Cancer

The nonrandomness of chromosome changes during tumor development has also been convincingly demonstrated in experimentally induced animal cancers. The Levan group used two different agents to induce tumors, Rous sarcoma virus (RSV) and 7–12-dimethylbenz(α)anthracene (DMBA). In mice, rabbits, Chinese hamsters, and rats, comparable results were obtained (cf. Levan et al, 1977). Let us consider the results of the rat experiments. Primary sarcomas induced by either RSV or DMBA start out with the normal chromosome complement of the rat. Gradually cells with abnormal chromosome constitutions appear and undergo a sequence of apparently predetermined chromosome changes, each agent inducing a specific nonrandom evolutionary pattern of its own. The origin of the tissue does not seem to affect the chromosomal development of the tumor.

In the RSV-induced tumors, the cells first gain an extra 7 chromosome, then a 13 chromosome, and finally a 12 chromosome. The pattern of evolution in the DMBA-induced sarcomas, on the other hand, is different. First an extra chromosome 2 is added and then one of the small metacentric chromosomes. Similar experiments are discussed by DiPaolo and Popescu (1976).

Mitotic Aberrations in Cancer Cells

By the end of the last century it was known that multipolar mitoses are common in human cancer. Since then, evidence of a great variety of mitotic aberrations has been accumulated. Chromosome breaks and structural rearrangements are also much more frequent in malignant than in normal cells (cf. Shaw and Chen, 1974). The cytological picture in an advanced cancer is so confused, ranging from small to giant nuclei (Fig.

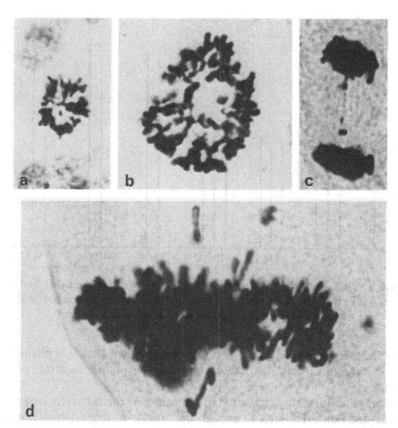

Figure XXVII.2. Mitotic stages from normal human placenta (a) and cervical cancer. (a) Diploid metaphase (polar view); (b) metaphase in octoploid range (polar view); (c) anaphase with a bridge; (d) giant metaphase (side view) (Therman et al, 1984).

XII.2) and small to enormous metaphases (Fig. XXVII.2), that only by studying a whole range of tumors from the early stages to highly anaplastic ones can one obtain an idea of the order in which the different aberrations appear and what their mutual relationships are. A series of such analyses of tumors ranging from early to highly malignant was performed around 1950 on primary human cancers of the female genital tract (cf. Timonen and Therman, 1950) and has been continued more recently (cf. Therman et al, 1983; Therman et al, 1984). These studies have shown that, in normal human tissues, prophase and metaphase require about the same time. This fact is reflected in their equal frequencies in counts on fixed biopsies; the ratio of metaphases to prophases is approximately 1 in normal tissues. In malignant tumors the ratio is increased and becomes higher in relation to the growing malignancy of the tumor, finally showing values of 23 to 35 and in some exceptional cases values as high as 50 (Fig.

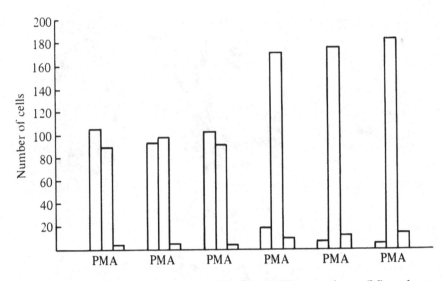

Figure XXVII.3. Relative frequencies of prophases (P), metaphases (M), and ana-phases (A) in three biopsies of normal epithelium of human fetal tubes (left) and three cases of cancer in adult fallopian tubes (data from Lehto, 1963).

XXVII.3). Recently Sisken et al (1982) have shown that this phenome-non is caused, at least in part, by an increase in the duration of the metaphases.

Another abnormality characteristic of cancer is the occurrence of mul-tipolar divisions. The first to appear are tripolar mitoses, which are fol-lowed by divisions with higher numbers of poles (Fig. XXVII.4) (Ther-man and Timonen, 1950). In addition to these basic phenomena, cancer cells display an almost infinite variety of other mitotic aberrations, such as endoreduplication, endomitosis, C-mitosis, lagging chromosomes, and chromatid bridges, as well as restitution at various stages (Chapter XII). Some of these may be secondary results of abnormal tumor physiology, especially anoxia (lack of oxygen). These observations have been made repeatedly in different types of human cancers, as well as in transplanta-ble and induced mouse tumors (cf. Oksala and Therman, 1974).

The mitotic aberrations in malignant cells, especially the increased metaphase/prophase ratio and the occurrence of multipolar divisions, are so characteristic of cancer that they have been used successfully in diag-nosis (Timonen, 1955).

Effect of Chromosome Changes on Tumor Development

The primary chromosome aberrations, which are characteristic of differ-ent types of malignant disease, are probably related to carcinogenesis itself. However, the striking changes in chromosome number and struc-

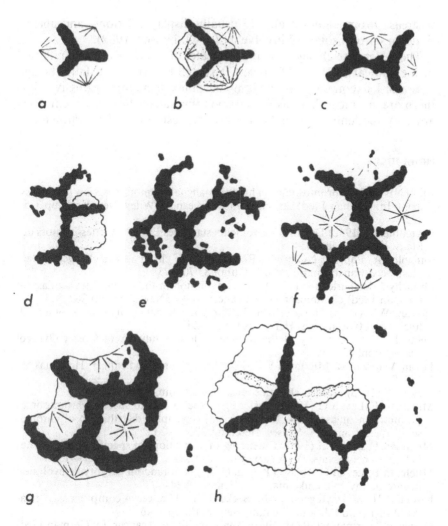

Figure XXVII.4. Multipolar metaphases from human cancer cells. (a) Tripolar metaphase; (b–c) quadripolar metaphases; (d–e) hexapolar metaphases; (f–g) pentapolar metaphases; (h) heptapolar metaphase (Therman and Timonen, 1950).

ture observed in most cancers take place during tumor progression. The mechanisms creating new chromosome constitutions are mitotic abnormalities together with chromosome structural changes. The result is a variety of chromosomally different cell types on which selection acts to promote the fastest dividing ones to form new stemlines. This also may explain the nonrandomness of the changes, since monosomy or trisomy for only certain chromosomes or chromosome segments will render the cells more malignant. For instance, chromosome 8 is, as a rule, trisomic, and chromosome 7 often monosomic. Trisomy for 1q is common, whereas chromosomes 5 and 20 usually display deletions, and aberrations for chromosome 17 involve i(17q) (cf. Nowell, 1982).

These stepwise changes leading to greater malignancy are also reflected in the clinical behavior of cancer. It is well-known that a fairly benign tumor or leukemia may at one jump become much more aggressive. Furthermore, a cancer that has been responding satisfactorily to chemotherapy may suddenly flare up, having become resistant to the treatment.

References

Atkin NB (1974) Chromosomes in human malignant tumors: a review and assessment. In: German J (ed) Chromosomes and cancer. Wiley, New York, pp 375–422

Bayreuther K (1952) Der Chromosomenbestand des Ehrlich-Ascites-Tumors der Maus. Naturforsch 7:554–557

DiPaolo JA, Popescu NC (1976) Relationship of chromosome changes to neoplastic cell transformation. Am J Pathol 85:709–738

Grouchy J de, Turleau C (1974) Clonal evolution in the myeloid leukemias. In: German J (ed) Chromosomes and cancer. Wiley, New York, pp 287–311

Heneen WK (1976) HeLa cells and their possible contamination of other cell lines: karyotype studies. Hereditas 82:217–247

Lehto L (1963) Cytology of the human Fallopian tube. Acta Obstet Gynecol Scand Suppl 42:1–95

Levan A, Levan G, Mitelman F (1977) Chromosomes and cancer. Hereditas 86: 15–30

Makino S (1975) Human chromosomes. Igaku Shoin, Tokyo

Mitelman F, Levan G (1978) Clustering of aberrations to specific chromosomes in human neoplasms. III. Incidence and geographic distribution of chromosome aberrations in 856 cases. Hereditas 89:207–232

Mitelman F, Levan G (1981) Clustering of aberrations to specific chromosomes in human neoplasms. IV. A survey of 1,871 cases. Hereditas 95:79–139

Mitelman F, Levan G, Nilsson P, et al (1976) Non-random karyotypic evolution in chronic myeloid leukemia. Int J Cancer 18:24–30

Nowell PC (1982) Cytogenetics. In: Becker FF (ed) Cancer: A comprehensive treatise, Vol 1 (2nd Edition). Plenum, New York, pp 3–46

Oksala T, Therman E (1974) Mitotic abnormalities and cancer. In: German J (ed) Chromosomes and cancer. Wiley, New York, pp 239–263

Rowley JD (1977) A possible role for nonrandom chromosomal changes in human hematologic malignancies. In: de la Chapelle A, Sorsa M (eds) Chromosomes today. Vol 6. Elsevier North-Holland, Amsterdam, pp 345–355

Sandberg AA (1979) The chromosomes in human cancer and leukemia. Elsevier North-Holland, New York

Sandberg AA, Sakurai M (1974) Chromosomes in the causation and progression of cancer and leukemia. In: Busch H (ed) The molecular biology of cancer. Academic, New York, pp 81–106

Sandberg AA, Wake N, Kohno S (1982) Chromosomes and causation of human cancer and leukemia. XLVII. Severe hypodiploidy and chromosome conglomerations in ALL. Cancer Genet Cytogenet 5:293–307

Shaw MW, Chen TR (1974) The application of banding techniques to tumor chromosomes. In: German J (ed) Chromosomes and cancer. Wiley, New York, pp 135–150

Sisken JE, Bonner SV, Grasch SD (1982) The prolongation of mitotic stages in SV-40-transformed vs nontransformed human fibroblast cells. J Cell Physiol 113:219–223

Therman E, Buchler DA, Nieminen U, et al (1984) Mitotic modifications and aberrations in human cervical cancer. Cancer Genet Cytogenet 11:185–197

Therman E, Sarto GE, Buchler DA (1983) The structure and origin of giant nuclei in human cancer cells. Cancer Genet Cytogenet 9:9–18

Therman E, Timonen S (1950) Multipolar spindles in human cancer cells. Hereditas 36:393–405

Timonen S (1955) Prophase index in the diagnosis of gynecological cancer. Ann Chir Gyn Fenn 4:222–233

Timonen S, Therman E (1950) The changes in the mitotic mechanism of human cancer cells. Cancer Res 10:431–439

Wolf U (1974) Theodor Boveri and his book "On the problem of the origin of malignant tumors." In: German J (ed) Chromosomes and cancer. Wiley, New York, pp 3–20

Yosida TH (1975) Chromosomal alterations and development of experimental tumors. In: Handbuch der allgemeinen Pathologie VI. Springer, Heidelberg, pp 677–753

Yunis JJ (1984) Clinical significance of high resolution chromosomes in the study of acute leukemias and non-Hodgkins lymphomas. In: Fairbanks VF (ed) Current hematology, Vol 3. Wiley, New York, pp 353–391

XXVIII
Mapping of Human Chromosomes

Gene Mapping

Mapping genes on the chromosomes is one of the fastest growing disciplines in human genetics. Reports on new assignments of genes to particular chromosomes, as well as books and reviews dealing with various aspects of chromosome mapping, appear regularly (cf. Ruddle and Creagan, 1975; Creagan and Ruddle, 1977; McKusick and Ruddle, 1977; Shows 1978; McKusick, 1983). In addition, the developments in this field have been reviewed almost every year at international meetings: New Haven in 1973; Rotterdam, 1974; Baltimore, 1975; Winnipeg, 1977; Edinburgh, 1979; Oslo, 1981; Los Angeles, 1983; and Helsinki, 1985. The results from the Los Angeles conference are published in Human Gene Mapping 7 (1984).

The considerable advances in gene mapping provide a good example of the successful application of different approaches to the same problem. These range from family studies to somatic cell genetics and from cytogenetics to biochemical and molecular studies. Often a gene assignment based on one technique has been confirmed with another. The traits for which loci have been mapped include inherited diseases, enzymes, serum and other proteins, surface and blood group antigens, RNA markers, susceptibility to drugs and toxins, viral markers, gene-regulating markers, blood-clotting factors (cf. Shows, 1978) and lately several oncogenes (cf. de la Chapelle and Berger, 1984).

In gene mapping as in many other branches of cytogenetics, chromosome banding provided a breakthrough. Genes can be assigned not only to individual chromosomes, but also to specific chromosome regions, and the accuracy of gene localization has been further improved by the use of prophase banding.

Gene assignments are classified as *confirmed* when at least two studies have come to the same conclusion; *provisional* when only one determination exists; and *controversial* when different studies provide contradictory assignments (cf. Donald and Hamerton, 1978).

Family Studies

It is usually easy to decide, on the basis of family studies, whether a gene lies on an autosome or on one of the sex chromosomes. An X-linked gene is never inherited from father to son, whereas a Y-linked gene always is, and cannot be passed from father to daughter. The first human gene to be assigned to a specific chromosome was red-green color blindness, which was assigned to the X chromosome in 1911 (cf. McKusick and Ruddle, 1977). Now more than 200 X-linked genes are known (cf. Miller et al, 1984). Very few genes have been assigned to the Y chromosome, the most important being the regulator of H-Y antigen on Yp (cf. Miller et al, 1984). Studies of linkage and crossing-over show whether or not two or more genes are on the same chromosome (*syntenic*), and what their relative distances are.

Marker Chromosomes

The assignment of a gene or a linkage group to a specific chromosome can be done on the basis of family studies only if a suitable marker chromosome is available. Typical markers consist of heterochromatic variants, fragile regions, or structurally abnormal chromosomes. The first gene assigned by means of a marker chromosome was the Duffy blood group locus, which segregated with a large variant of the centric heterochromatin in chromosome 1 (Donahue et al, 1968). The gene for alpha-haptoglobin was linked to a fragile site on 16q (such a fragile site is illustrated in Fig. VII.5a and b) in 30 family members and was separated from it in three (Magenis et al, 1970). This finding indicates that the alpha-Hp locus lies near the fragile site on 16q. A factor causing mental retardation follows the fragile region near the distal end of Xq (Fig. XX.4d) (cf. Howard-Peebles and Stoddard, 1979).

The major histocompatibility complex (HLA) was assigned to chromosome 6 by means of a pericentric inversion segregating in a family (Lamm et al, 1974). Reciprocal translocations involving chromosome 6 made the further localization of this gene to 6p21 possible (Breuning et al, 1977; Francke et al, 1977). This assignment was, in turn, confirmed through a study of partial trisomy for the segment distal to 6p21 that resulted from crossing-over in a pericentric inversion (Pearson et al, 1979).

Cell Hybridization

Although refinements in banding techniques have made more and more chromosome markers available (it is claimed that by now every human being can be distinguished by them), gene mapping owes its greatest advances to other methods, especially cell hybridization. All cell hybridization techniques are based on the observation that somatic cells of the same species or of two different species will fuse under certain conditions. For purposes of human chromosome mapping, a human and a rodent cell are usually hybridized. Mouse cells are often used as the nonhuman parent strain. The following criteria should be considered when the cell types are chosen for human gene mapping (cf. Creagan and Ruddle, 1977):

1) The cells grow rapidly in culture.
2) The cells are easy to hybridize, and the hybrid cells divide in culture.
3) The chromosomes of the parent cells can be identified without difficulty (Fig. XXVIII.1).

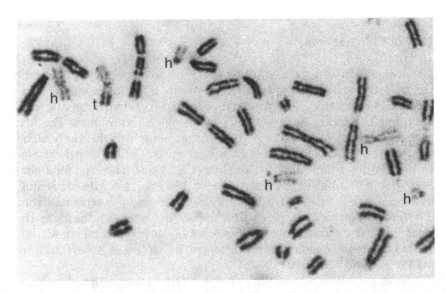

Figure XXVIII.1. Part of the metaphase plate from a man–mouse hybrid cell stained with G-11. Mouse chromosomes (aneuploid cell line) are dark with light centric regions, human chromosomes (h) are light with dark centric regions, and translocation chromosome (t) between the two species is part dark, part light (courtesy of RI DeMars).

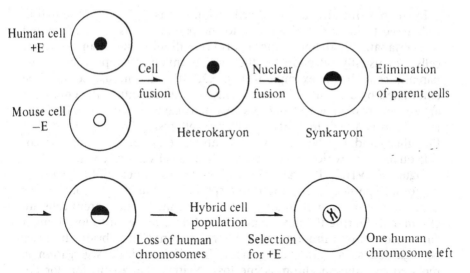

Figure XXVIII.2. Assignment of the gene coding for the enzyme E to a specific human chromosome (black) through fusion of a human cell (nucleus black) with a mouse cell (nucleus outlined) followed by cell selection.

4) Human chromosomes are unilaterally lost from the hybrid cells (see following paragraphs).
5) The human phenotypic markers can be easily determined and are distinguishable from those of the other parent cell.

Spontaneous fusion of somatic cells is an extremely rare phenomenon, which probably occurs only when one cell type is malignant. (An exception is provided by cells of Bloom's syndrome patients, as described in Chapter XI). The incidence of cell fusions can be increased considerably by treating cultured cells with inactivated Sendai virus or a chemical agent, such as polyethylene glycol (cf. Ruddle and Kucherlapati, 1974). When two cells fuse, the hybrid at first has two nuclei; it is a *heterokaryon*. After nuclear fusion, the cell is called a *synkaryon* (Fig. XXVIII.2).

Selection

Different selection systems play an important role in all cell hybridization methods. They include selection for a specific gene (or, more accurately, for the phenotype it causes), for hybrid cells against the background of parental cells, or for a specific chromosome or chromosome segment. Selection may be positive or negative, relative or absolute. The means used include culture media in which a certain cell type cannot grow or grows preferentially, or selective killing of one cell type with toxins, antibodies, or viruses to which the other cells are resistant.

Even an artificially increased cell fusion rate is so low that the hybrid cells have to be enriched relative to the parental cells (cf. Ruddle and Kucherlapati, 1974). Sometimes hybrid cells divide faster than the parent cells, but usually selection for the hybrids involves suppression of the parental cells. In the example in Fig. XXVIII.2, the mouse cells can be eliminated by a medium in which only cells that produce the enzyme E are able to grow. (The cell markers are often enzymes, which can be determined through their mobility in gel electrophoresis.) The human cells, on the other hand, can be killed with diphtheria toxin or the cardiac glycoside ouabain, to which the mouse and hybrid cells are insensitive.

Figure XXVIII.2 illustrates the steps in the assignment of the gene coding for a hypothetical enzyme E to a specific human chromosome. After the fusion of the human and mouse cells and nuclei, the parent cells are eliminated. During the next phase the hybrid cells begin to lose human chromosomes in a more or less random fashion. Cell hybridization can be regarded as a parasexual process in which meiotic segregation is replaced by random chromosome loss. Neither the reason for nor the exact mechanism of the process of chromosome elimination is understood, although it plays an important role in the gene mapping procedures.

Chromosome elimination leads to the formation of hybrid clones that differ in their human chromosome content. At this point, selection for or against a certain phenotype in the hybrid cells is an important tool in the gene mapping process. In our example (Fig. XXVIII.2), a selection system is used that allows only those cells producing enzyme E to survive. Naturally the clearest result is obtained when cells with phenotype E have only one human chromosome left, which must therefore be the site of gene E. The first human enzyme to be assigned in this way was thymidine kinase (TK) to chromosome 17 (cf. Ruddle and Creagan, 1975).

The usual result of random chromosome elimination is a collection of clones with different combinations of human chromosomes. A selected *clone panel* may enable the investigator to assign a gene to a specific chromosome. If all the clones displaying a particular phenotype have one and only one human chromosome in common, the conclusion is inevitable that this one is the site of the gene (cf. Ruddle and Creagan, 1975). With the clone panel technique it is also possible to determine whether or not two or more genes are syntenic.

Chromosome Translocations

The order of genes on a chromosome, as well as their assignment to a specific chromosome segment, can be determined by means of reciprocal translocations involving the relevant chromosome. The co-occurrence in a hybrid cell of a particular phenotype accompanied by one of the translocation chromosomes restricts the location of the gene to the part of the chromosome involved. By the use of several, partly overlapping translocations, the assignment of a gene can be limited to a smaller and smaller chromosome segment.

This was done for three X-linked genes by means of a translocation in which almost the whole Xq was attached to the distal end of 14q (Allderdice et al, 1978). Cell hybridization studies showed that the genes for HGPRT (hypoxanthine-guanine phosphoribosyl-transferase), PGK (phosphoglycerate kinase), and G6PD (glucose–6-phosphate dehydrogenase) segregated with the long translocation chromosome as did the autosomal gene NP (nucleoside phosphorylase). The three X-linked genes could thus be assigned to Xq and the autosomal gene to 14q (Ricciuti and Ruddle, 1973). By means of other X-autosomal translocations it was possible to demonstrate that the order of the three genes was: centromere-PGK-HGPRT-G6PD and to determine the limits of the segments within which each of them must be situated.

Two different gene maps have been developed for many human chromosomes. The one type of map contains the genes whose order and relative distances have been determined with family studies, whereas in the other type the genes that can be studied on the cellular level are assigned to specific segments (cf. McKusick and Ruddle, 1977; Shows, 1978). Attempts have been made to bring the two maps together (cf. Francke et al, 1977).

The enzyme TK, already mentioned, which had been assigned to chromosome 17, was found to lie on 17q when this arm was translocated to a mouse chromosome in a hybrid cell (cf. Ruddle and Kucherlapati, 1974). This was the first example of a translocation between two chromosomes of different species. An even more accurate assignment to the segment distal to 17q21 was possible with the help of a translocation between two human chromosomes (Yoshida and Matsuya, 1976).

Other Uses of Deletions and Translocations in Mapping

Recently a number of single genes (or in some cases syndromes apparently caused by a few genes) have been mapped using prophase banding to determine deletions or translocation break points. Thus the gene for Duchenne muscular dystrophy has been mapped to Xp21, when in several affected females the translocation break point coincided with this band, and a few other X-linked genes have been assigned in the same way (cf. Riccardi, 1984). Wilms' tumor-aniridia syndrome has been found to be caused by a deletion of 11p13. On the other hand, the connection of the Prader-Willi syndrome and 15q11-q13 is less clear, since the patients have shown deletions, duplications, or translocations involving this region (Ledbetter et al, 1982). As discussed in Chapter XXVI, retinoblastoma seems to be correlated with the hemizygosity of 13q14.

Transfer of Microcells and Single Chromosomes

A refinement of the cell hybridization techniques is the fusion of a "microcell" with a normal cell. A microcell consists of a nucleus with a few (or even a single) chromosomes surrounded by a small amount of cytoplasm. Microcells are prepared by treating normal cells with colchicine to scatter the chromosomes around the cell and then treating the cell with cytochalasin B to break it up. The microcell can then be fused in the usual way with a complete recipient cell (cf. Fournier and Ruddle, 1977).

Another modification of the cell hybridization technique is the transfer of a single metaphase chromosome into a recipient cell. The transferred chromosome falls into fragments in the alien cytoplasm, leaving the gene under study attached to a segment of DNA. Such hybrid cells are unstable until, many cell generations later, the fragment may attach itself to a host chromosome.

Gene Dosage

Mapping by means of gene dosage is either qualitative or quantitative. In the former case, the number of different gene products is compared with the number of genes present. Quantitative studies, on the other hand, correlate the number of alleles with the quantity of a gene product (cf. Creagan and Ruddle, 1977). Both types of studies take advantage of chromosome deletions and duplications (partial monosomy or trisomy). No primary gene assignment has been made solely by means of trisomy mapping, but a number of previous assignments have been confirmed. A deletion of the distal end of 2p enabled the mapping of the gene for acid phosphatase to the missing segment (cf. Aitken et al, 1976). Several other assignments have been confirmed with deletion mapping (cf. McKusick and Ruddle, 1977).

Exclusion mapping is the reverse of trisomy and monosomy mapping. This method determines the genes that are *not* affected by the duplication or deletion of a chromosome segment. A study of 20 deletions made possible the exclusion of an average of four genes from each deleted segment (cf. Aitken et al, 1976).

In-situ Hybridization

When a single-stranded probe of DNA or complementary RNA is labeled with a radioactive isotope and hybridized to a chromosome preparation in which the chromosomes have been treated to separate the DNA strands, the original site of the DNA is marked with silver grains after autoradiography (cf. Pardue and Gall, 1975). In this way the human satellite DNAs have been localized to the centric and Y heterochromatin (cf.

Yunis et al, 1977). The repeated genes, which occur in 100 to 200 copies in the human chromosome complement, have also been mapped by direct hybridization (cf. Evans and Atwood, 1978). They include the genes coding for the ribosomal proteins 18S and 28S, which are situated on the satellite stalks of acrocentric chromosomes. The genes producing the 5S ribosomal protein have been assigned to 1q42-q43 (cf. Steffensen, 1977). The assignment of the histone genes, repeated about 40 times, to 7q32-q36 was confirmed by direct hybridization on prophase-banded chromosomes (Chandler et al, 1979).

Single genes were first localized with in-situ hybridization to the poly-tene chromosomes of *Drosophila*, in which a band may consist of thousands of copies of a gene (Fig. XXVIII.3). The first human single genes were localized with this technique in 1981. These were alpha-globin genes assigned to chromosome 16 (Gerhard et al, 1981) and beta-globin genes to 11p (Malcolm et al, 1981).

The in-situ hybridization has opened up completely new possibilities for gene mapping. Any gene or DNA sequence that has been cloned can now be mapped (cf. Harper and Saunders, 1984). The great advantage of

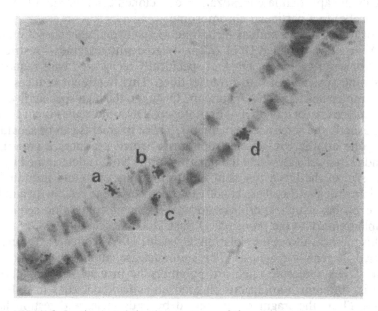

Figure XXVIII.3. In-situ hybridization to *Drosophila* polytene chromosomes. The chromosomes were hybridized with ³H-labeled RNA homologous to a *Drosophila* transposable element known as the *P* element and autoradiographed. The silver grains reveal the chromosomal sites of insertion of this element. At several sites (a and c) it can be seen that only one of the paired homologues contains an insert. Site c is the location of a new mutation caused in this instance by insertion there of a *P* element (courtesy of K Loughney and B Ganetzky).

this technique, apart from accuracy, is that a gene need not be expressed, which is a prerequisite for gene mapping with the cell hybridization technique. Single gene mapping differs from the assignment of longer DNA sequences in that information from a number of autoradiographed metaphases has to be pooled to localize the gene to the region that contains the highest total number of silver grains.

Skolnick et al (1984) report that of 132 cloned human genes, 116 have been mapped to specific chromosomes. In addition to the alpha- and beta- globin genes, for instance, the following important genes have been mapped with in-situ hybridization: those responsible for the immunoglobulins, interferon, insulin, collagen type I alpha-1, growth hormone, myosin heavy- chain, as well as several oncogenes.

Recombinant DNA Gene Mapping Methods

In-situ hybridization involves the interaction between cloned (or otherwise purified) DNA (or complementary RNA transcripts thereof) as a probe and the DNA of intact chromosomes as target. The technique enables the approximate localization of a cloned gene (or any other specific probe) to a particular chromosome.

Biochemists use a similar technique to achieve a higher resolution of DNA structure. Isolated DNA can be treated with enzymes—restriction endonucleases—that cut DNA at specifically recognized base sequences consisting typically of 4 to 6 nucleotides. This treatment reduces large DNA molecules to a set of fragments. Owing to the high specificity of the endonucleases, each of a set of homologous DNA molecules is cut to produce exactly the same set of fragments. These fragments can be separated from one another by gel electrophoresis, which produces a pattern of "bands" in a gel slab where the position of the band is determined by the exact size of the DNA fragment. Following transfer to and immobilization in nitrocellulose, these bands can then be made to react with a radioactive probe, which hybridizes to any fragment containing sequences complementary to the probe (cf. D'Eustachio and Ruddle, 1983).

This method allows the isolation of specific DNA sequences. It can also be used in genetic diagnosis. The endonuclease cut sites on a chromosomal DNA molecule often are polymorphic in a human population. That is, different homologous chromosomes have cut sites in different places. Thus, the fragments produced by endonuclease digestion have characteristically different sizes, and these size differences will be revealed by the technique just described. If the fragment size class of the polymorphism contains a mutant gene, the fragment size can be used as a marker showing linkage to the gene (cf. Epstein et al, 1983).

A further procedure that has also found use in gene mapping is the isolation and sorting of chromosomes with a fluorescence-activated cell sorter. Each fraction, which consists of one to three chromosome types, can be purified further and used to obtain DNA from a specific chromosome (cf. Lebo, 1982; Yu et al, 1984).

Conclusions

Constitutive heterochromatin is naturally devoid of any mendelian genes, and there is also convincing evidence that the Q-bright chromosome bands consist to a considerable extent of intercalarary heterochromatin and therefore contain relatively few genes. In contrast, the Q-dark regions have a high gene density, and important genes are apparently concentrated in certain hot spots (Chapter V).

The total number of human genes is still a matter of guesswork (for instance, the haploid *Drosophila* genome is believed to have 5,000–6,000 genes), and obviously only a fraction of them are known. McKusick (1983) lists some 1600 autosomal genes, of which about a third are mapped to specific chromosomes. Because of the easy detection of X-linkage, more than 200 genes are known to be on this chromosome.

More than half of the localized genes have been mapped by means of cell hybridization. However, it seems safe to assume that the future belongs to in-situ hybridization and other molecular approaches. We should, however, remember that, as so often before, techniques as yet unknown may revolutionize gene mapping in the years to come.

One of the goals of human genetics is the development of a complete gene map. This accomplishment lies far in the future, since the overwhelming majority of the genes are still unknown. One of the tasks ahead is the coordination of the results of formal genetics, somatic cell genetics, and molecular studies on chromosome structure. The crowning achievement at the molecular level will be the base sequencing of the total DNA of the human genome and the correlation of such data with the functions of different parts of the chromosomes.

References

Aitken DA, Ferguson-Smith MA, Dick HM (1976) Gene mapping by exclusion: the current status. Cytogenet Cell Genet 16:256–265

Allderdice PW, Miller OJ, Miller DA, et al (1978) Spreading of inactivation in an (X; 14) translocation. Am J Med Genet 2:233–240

Breuning MH, Berg-Loonen EM van den, Bernini LF, et al (1977) Localization of HLA on the short arm of chromosome 6. Hum Genet 37:131-139

Chandler ME, Kedes LH, Cohn RH, et al (1979) Genes coding for histone proteins in man are located on the distal end of the long arm of chromosome 7. Science 205:908–910

Chapelle A de la, Berger R (1984) Report of the committee on chromosome rearrangements in neoplasia and on fragile sites. Cytogenet Cell Genet 37:274–311

Creagan RP, Ruddle FH (1977) New approaches to human gene mapping by somatic cell genetics. In: Yunis JJ (ed) Molecular structure of human chromosomes. Academic, New York, pp 89-142

D'Eustachio P, Ruddle FH (1983) Somatic cell genetics and gene families. Science 220:919-924

Donahue RP, Bias WB, Renwick JH, et al (1968) Probable assignment of the Duffy blood group locus to chromosome 1 in man. Proc Natl Acad Sci USA 61:949-955

Donald LJ, Hamerton JL (1978) A summary of the human gene map, 1973-1977. Cytogenet Cell Genet 22:5-11

Epstein CJ, Cox DR, Schonberg SA, et al (1983) Recent developments in the prenatal diagnosis of genetic diseases and birth defects. Annu Rev Genet 17:49-83

Evans HJ, Atwood KC (1978) Report of the committee on in situ hybridization. Cytogenet Cell Genet 22:146-149

Fournier REK, Ruddle FH (1977) Microcell-mediated chromosome transfer. In: Sparkes RS, Comings DE, Fox CF (eds). Molecular human cytogenetics. Academic, New York, pp 189-199

Francke U, George DL, Pellegrino MA (1977) Regional mapping of gene loci on human chromosomes 1 and 6 by interspecific hybridization of cells with a t(1;6)(p3200;p2100) translocation and by correlation with linkage data. In: Sparkes RS, Comings DE, Fox CF (eds). Molecular human cytogenetics. Academic, New York, pp 201-216

Gerhard DS, Kawasaki ES, Bancroft FC, et al (1981) Localization of a unique gene by direct hybridization *in situ*. Proc Natl Acad Sci USA 78:3755- 3759

Harper ME, Saunders GF (1984) Localization of single-copy genes on human chromosomes by in situ hybridization of ^{3}H-probes and autoradiography. In: Sparkes RS, de la Cruz FF (eds). Research perspectives in cytogenetics. University Park Press, Baltimore, pp 117-133

Howard-Peebles PN, Stoddard GR (1979) X-linked mental retardation with macro-orchidism and marker X chromosomes. Hum Genet 50:247-251

Human gene mapping 7 (1984) Cytogenet Cell Genet 37, Nos 1-4

Lamm LU, Friedrich U, Petersen GB, et al (1974) Assignment of the major histocompatibility complex to chromosome no. 6 in a family with a pericentric inversion. Hum Hered 24:273-284

Lebo RV (1982) Chromosome sorting and DNA sequence localization. Cytometry 3:145-154

Ledbetter DH, Mascarello JT, Riccardi VM, et al (1982) Chromosome 15 abnormalities and the Prader-Willi syndrome: A follow-up report of 40 cases. Am J Hum Genet 34:278-285

Magenis RE, Hecht F, Lovrien EW (1970) Heritable fragile site on chromosome 16: probable localization of haptoglobin locus in man. Science 170:85- 87

Malcolm S, Barton P, Murphy C, et al (1981) Chromosomal localization of a single copy gene by *in situ* hybridization - human beta-globin genes on the short arm of chromosome 11. Ann Hum Genet 45:135-141

McKusick VA (1983) Mendelian inheritance in man, 6th edn. Johns Hopkins University Press, Baltimore

McKusick VA, Ruddle FH (1977) The status of the gene map of the human chromosomes. Science 196:390-405

Miller OJ, Drayna D, Goodfellow P (1984) Report of the committee on the genetic constitution of the X and Y chromosomes. Cytogenet Cell Genet 37:176-204

Pardue ML, Gall JG (1975) Nucleic acid hybridization to the DNA of cytological preparations. In: Prescott DM (ed) Methods in cell biology 10. Academic, New York, pp 1-16

Pearson G, Mann JD, Bensen J, et al (1979) Inversion duplication of chromosome 6 with trisomic codominant expression of HLA antigens. Am J Hum Genet 31:30-34

Riccardi VM (1984) High resolution karyotype-phenotype correlations and focused chromosome analysis. In: Sparkes RS, de la Cruz FF (eds). Research perspectives in cytogenetics, University Park Press, Baltimore, pp 53-62

Ricciuti FC, Ruddle FH (1973) Assignment of three gene loci (PGK, HGPRT, G6PD) to the long arm of the human X chromosome by somatic cell genetics. Genetics 74:661-678

Ruddle FH, Creagan RP (1975) Parasexual approaches to the genetics of man. Annu Rev Genet 9:407-486

Ruddle FH, Kucherlapati RS (1974) Hybrid cells and human genes. Sci Am 231:36-44

Shows TB (1978) Mapping the human genome and metabolic diseases. Birth Defects, Proc 5th Internat Conf: 66-84

Skolnick MH, Willard HF, Menlove LA (1984) Report of the committee on human gene mapping by recombinant DNA techniques. Cytogenet Cell Genet 37:210-273

Steffensen DM (1977) Human gene localization by RNA:DNA hybridization *in situ*. In: Yunis JJ (ed) Molecular structure of human chromosomes. Academic, New York, pp 59-88

Yoshida MC, Matsuya Y (1976) Confirmation of the human thymidine kinase locus, 17q21→17qter, by means of a man-mouse somatic cell hybrid, D98/AH-2 X LMTK⁻ Cl-1D. Hum Genet 31:235-239

Yu L-C, Gray ZW, Langlois R, et al (1984) Human chromosome karyotyping and molecular biology by flow cytometry, In: Sparkes RS, de la Cruz FF (eds). Research perspectives in cytogenetics. University Park Press, Baltimore, pp 63-73

Yunis JJ, Tsai MY, Willey AM (1977) Molecular organization and function of the human genome. In: Yunis JJ (ed) Molecular structure of human chromosomes. Academic, New York, pp 1-33

Author Index

Abe, 247
Adler, 83
Aitken, 288
Alberts, 58–59
Alitalo, 268
Allderdice, 287
Alter, 89
Alvesalo, 164, 173
Ambros, 96
André, 223
Andrle, 203
Arrighi, 36–37
Atkin, 103, 274
Atkins, 245
Auerbach, 66, 80
Aurias, 229
Awa, 83
Aymé, 155

Bahr, 258
Balaban-Malenbaum, 258, 260
Balkan, 239, 252
Barker, 258
Barr, 4, 166
Bartram, 65
Bartsch-Sandhoff, 250
Bass, 235
Bateman, 70
Bault, de, 206

Bayreuther, 2, 274
Beçak, 23
Beerman, 115, 122–123
Benirschke, 203
Bennett, 154
Berghe, van den, 220
Biederman, 223, 231
Biedler, 256–259
Bijlsma, 79, 230
Bishop, 264
Bloom, 37
Bochkov, 78
Bojko, 138
Borgaonkar, 198, 228, 232
Bosman, 92
Boué, A., 194, 196, 198, 206
Boué, J., 54, 194, 198, 207
Boveri, 273
Brandriff, 35, 152
Breuning, 283
Brown, S. W., 50
Brown, W. T., 92
Bruun Petersen, 92
Bull, 161
Burnham, 197, 214, 228

Camargo, 55
Carpenter, A. T. C., 141
Carpenter, N. J., 187, 190

Carr, 155, 178, 194–195, 198, 206
Caspersson, 3, 21, 36, 250
Cattanach, 171–172
Cavenee, 270
Centerwall, 234
Chaganti, 89–91, 97
Chandler, 289
Chandley, 153
Chapelle, de la, 75, 173, 265, 267, 282
Chapman, 247, 251
Christensen, 222
Cohen, 70, 160, 162, 245
Coldwell, 206
Coleman, 79
Cooke, 69
Cooper, 265, 271
Cowell, 256, 258–259
Craig-Holmes, 213
Creagan, 103, 282, 284, 288
Cremer, 28, 107

Dallapiccola, 219
Daly, 71, 179–180, 188
D'Amato, 116, 119–120
Daniel, 162, 247
Darlington, 23, 32, 46, 127, 136, 138, 141, 152
Davidson, 171
Davis, 234
DeMars, 284
Denniston, 178
D'Eustachio, 290
DiPaolo, 276
Djalali, 215
Dodson, 196
Donahue, 283
Donald, 283
Doolittle, 52
Drets, 36, 41, 70, 72, 190, 224–225
Drewry, 51, 233
Drwinga, 107
Dudits, 105
Dutrillaux, 32, 35–36, 70, 161

Edwards, J. H., 3
Edwards, R. G., 34
Eiberg, 37

Epstein, 34, 290
Evans, H. J., 36, 54, 65–66, 80, 289
Evans, J. A., 250

Fankhauser, 195
Farah, 247
Fialkow, 271
Finnegan, 46
Fitzgerald, 72, 190, 230
Flavell, 52, 100
Ford, 2–3, 243
Forejt, 153
Fournier, 288
Fraccaro, 229, 237
Francke, 199, 270, 283, 287
Fried, 248
Friedrich, 164
Fryns, 74, 188
Funderburk, 230
Furbetta, 250

Gagnon, 156
Gall, 50, 58
Galloway, 180
Gartler, 171–172
Gatti, 91
Gebhart, 66, 80, 83
Geitler, 115–116, 119
Gerhard, 289
German, 88–89, 92, 97
Gilbert, 269, 271
Gillies, 140
Glover, 191
Goldman, 55
Goodpasture, 49
Gorlin, 201, 203–204
Gracias-Espinal, 250
Gropp, 200
Grouchy, de, 177, 198, 202–204, 211, 216, 228, 232, 276
Gustavii, 34
Gustavson, 92

Habedank, 237
Hagemeijer, 55
Hager, 212

Hamerton, 141, 177, 180, 198, 251
Hansmann, 69
Harnden, 82
Harper, 289
Harris, 252
Hassold, 154–155, 194
Hayflick, 78
Hecht, 75, 156, 219, 231, 248, 264
Heitz, 46, 49–50, 116, 119, 122
Hellkuhl, 187
Henderson, A. S., 49
Henderson, S. A., 155
Heneen, 274
Herva, 183
Hittelman, 107
Hodes, 202–203
Holm, 138, 141, 147
Holmquist, 70
Hongell, 247
Hoo, 220
Hood, 45–46
Hook, 179, 201, 238, 247
Hotta, 135
Howard-Peebles, 191, 283
Hsu, L. Y. F., 204
Hsu, T. C., 1–2, 4, 17, 50, 248
Hultén, 4, 34, 134, 144–146

Inouye, 220

Jackson, 16
Jacobs, 3, 157, 196, 232, 247
Jacobsen, 235
Jagiello, 34
John, 128, 138, 244
Johnson, 206
Jones, G. H., 138
Jones, K. W., 5
Jones, R. N., 52
Juberg, 155

Kahan, 172
Kaiser, 213–214, 252
Kaiser-McCaw, 91
Kajii, 157, 196, 237
Kallio, 123

Kanda, 96
Karp, 154
Kato, 97, 107
Kihlman, 65, 80, 154
Kiknadze, 115
Kirkels, 250
Kleczkowska, 221
Knudson, 271
Knuutila, 103
Korenberg, 38, 54, 94–95, 100, 207,
    232
Kornberg, 58
Koskull, von, 82, 248
Koulischer, 153
Kreber, 114
Kuhn, 36, 47, 54, 67, 73, 78, 89, 91,
    98, 100, 199, 232, 259, 270

Lambert, 225
Lamm, 283
Land, 264, 268, 271
Lande, 228
Lansky, 220–221
Latt, 4, 31, 36, 38, 94, 96–97, 169–
    170, 185
Lauritsen, 197–198, 207
Laxova, 217
Lebo, 291
Ledbetter, 287
Lehto, 278
Lejeune, 2, 203
Leonard, 214
Leschot, 223
Lester, 172
Levan, A., 1–2, 110, 116, 154, 258–
    260, 274, 276
Levan, G., 258
Lewandowski, 198–199
Lewin, 58
Lewis, 239
Lilienfeld, 203
Lindenbaum, 237, 239–240
Loughney, 289
Luciani, 141
Luthardt, 155
Luykx, 139
Lynch, 263
Lyon, 4, 170–171, 178

Madan, 213
Magenis, 213, 283
Magnelli, 161
Makino, 1–2, 5, 198, 274
Malcolm, 289
Mameli, 247
Maraschio, 183–184
Mark, 269
Martin, 35, 152
Marx, 269
Mascarello, 223
Matsubara, 221
Mattei, J. F., 155, 191
Mattei, M. G., 184, 186, 190
Mayer, 213
McClure, 203
McDermott, 151
McKusick, 282–283, 287–288, 291
Meer, 230
Mendelsohn, 41
Metaxotou, 205
Michels, 234
Migeon, B. R., 171, 173
Migeon, C. J., 173
Mikelsaar, 212, 245
Mikkelsen, 156, 244, 247–248, 251–252
Miklos, 45, 52
Miller, O. J., 48, 248, 259–260, 283
Miller, O. L., 61
Miller, R. W., 263
Mitelman, 275
Mittwoch, 170
Mohandas, 172
Moorhead, 3, 33, 92, 214
Motl, 67, 220–221
Müller, 163
Murken, 179
Murphree, 270

Nagl, 53, 110, 115–116, 119–123
Nakagome, 72, 206, 229
Nakajima, 220
Narahara, 189
Nichols, 78, 82, 84
Niebuhr, 69, 195, 201–203, 216–217, 219, 244
Nielsen, J., 178, 200, 244, 248

Nielsen, K. B., 223
Niikawa, 243, 250
Niss, 220
Noel, 203
Nowell, 3, 263, 268, 274, 280
Nuzzo, 234

Obe, 107–108
Ohno, 4–5
Oksala, 110, 112, 116–117, 121, 127, 131, 138, 252, 278
Opitz, 217
Orye, 245
Otto, 89, 103, 105–108, 188

Painter, 2, 119
Palmer, 230, 244, 250
Pantzar, 234
Pardue, 36, 288
Passarge, 46, 88
Patau, 3, 20–21, 29, 41, 53, 69, 97, 120, 150, 200–201, 207
Pathak, 34, 140–141
Patil, 206
Paton, 200
Pawsey, 92
Pearson, 284
Penrose, 203
Pérez-Castillo, 247
Pihko, 55, 229, 232, 234–235
Polani, 138, 178–179
Prieto, 230

Raimondi, 231
Rao, 104, 107
Rapp, 80, 151, 153
Redheendran, 201
Rees, 19, 138
Rhomberg, 221
Riccardi, 206, 287
Ricciuti, 287
Rieger, 65, 80, 82, 154
Riley, 221
Ris, 15, 36, 47, 53–54, 61, 139
Roberts, 186
Robertson, 268

Robinson, 161
Rockman-Greenberg, 245
Rodiere, 223
Röhrborn, 154
Rosenfeld, 162
Rothfels, 3
Rowley, 265, 267–268, 275
Rudak, 34, 152
Ruddle, 282, 285–287
Russell, 4, 83

Sanchez, 36, 40–41, 213
Sandberg, 96, 172, 256, 265, 274–275
Sanger, 176
Sarto, 116, 121, 170, 172–173, 187,
    196, 225, 231, 234, 245
Scarbrough, 196
Scheer, 60, 62–63
Schempp, 188
Schimke, 101, 259–261
Schinzel, 198, 204, 206, 211, 216, 228,
    232
Schmid, 162, 221
Schmidt, 55
Schnedl, 52, 221
Schober, 243
Schoeller, 84
Schröcksnadel, 195
Schroeder, H. J., 216
Schroeder, T. M., 48, 79, 91, 212
Schubert, 82–83, 96–97
Schuh, 250
Schwartz, 70
Schweizer, 37, 51
Seabright, 79
Searle, 153, 231
Sears, 152
Seuánez, 4–5
Shaw, D. D., 82
Shaw, M. W., 83, 276
Shepard, 154
Shows, 282, 287
Simoni, 34
Sirota, 176, 179
Sisken, 278
Skolnick, 290
Smith, 3, 198–199, 216
Smithies, 100

Solari, 34, 144, 147, 151
Solomon, 271
Sparrow, 65
Sperling, 27–28, 103–104, 107
Spowart, 189
Stahl, 4, 34, 63, 143
Stahl-Maugé, 73
Stebbins, 17–18, 24
Steffensen, 289
Stene, 155
Stern, 128, 135
Summitt, 187
Sunkara, 105
Surana, 221
Sutherland, 75

Takagi, 170, 172
Taylor, J. H., 29, 94
Taylor, M. C., 162
Tharapel, 212
Therkelsen, 222
Therman, 4–5, 14–15, 28, 34, 69, 71,
    79–80, 89, 97–99, 113, 115–117,
    120–123, 127–129, 131, 143, 152,
    156, 161, 182–189, 202, 231, 250,
    260, 270, 277–279
Tiepolo, 74
Timonen, 277–278
Tjio, 1–2, 33
Toomey, 222
Trunca, 48, 107, 183, 214–215, 222,
    229–230, 232, 238–240
Trunca Doyle, 213
Tschermak-Woess, 115, 119–120
Tuck, 156
Turleau, 203

Uchida, 154–157, 204, 216, 250
Ulber, 51

Valcárcel, 215
Vanderlyn, 46, 50
Vassilakos, 196
Verma, 212–213, 252
Vianna-Morgante, 215, 245
Vogel, F., 154

Vogel, W., 222
Vosa, 52

Wachtel, 162-163, 172-173,
   180
Wagenbichler, 155
Wahlström, 188
Walters, 128
Ward, 234
Watt, 230
Webb, 156
Weisblum, 53
Weitkamp, 73
Westergaard, 138-139
Wettstein, von, 138-139, 141
White, 28, 79, 138, 228
Whitehouse, 138
Wilson, 46, 128
Wisniewski, K., 206
Wisniewski, L., 223
Witkin, 179

Wolf, 162, 171, 173, 188, 273
Wolff, 94, 96
Wurster-Hill, 178
Wyss, 183, 187

Yamasaki, 47
Yoshida, 287
Yosida, 274
Yu, 41 291
Yunis, E., 205, 217
Yunis, J. J., 1, 3, 5, 32, 38, 40, 47, 54,
   75, 198, 264-269, 276, 289

Zackai, 219, 229, 237
Zakharov, 70, 94, 184, 190
Zankl, 69, 269
Zhou, 250
Žižka, 247
Zuffardi, 153, 187, 231, 248
Zybina, 115

Subject Index

Abortion, spontaneous, 10, 54, 100,
    154–157, 178, 194–198, 201,
    206–207, 233–240, 247–250
  heteroploid, 195
  induced, 34
Acentric fragment, 66–68, 228
Acentric ring, 66–67
Acetabularia, 63
Acid phosphatase gene, 288
Acridine orange, 38
Age
  chromosome breaks and, 78–79, 88
  maternal, nondisjunction and, 154–
    156
  paternal, nondisjunction and, 155
  sex chromosome abnormalities and,
    176
  spontaneous abortions and, 206
Agrobacterium tumefaciens, 263
Alkylating agents, 96
Allium, 138
Allocycly, 46, 107–108, 170, 172
alpha-Globin gene, 289
alpha-Haptoglobin gene, 283
Alzheimer's disease, 92
Amenorrhea, 183–184, 187
Amniocentesis, 33–34
Amphibians, 194–195
Amphiuma, 19

Anaphase, 14, 26–27, 126, 130–134,
    141
  in cancer cells, 277–278
Anemone blanda, 51
Aneuploidy, 23, 149, 155–156
  autosomal, 197
  in newborns, 238
  sex chromosome, with male
    phenotype, 179
  X chromosome, with female
    phenotype, 176–178
"Aneusomie de recombinaison", 214
Aniridia, 216, 266, 287
Anophthalmia, 201–202
Anotomys leander, 17
Antirrhinum, 197
Aphidicolin, 75
Arm ratio, 15
Ascites tumor, 274
Ataxia telangiectasia (AT), 88, 91, 96,
    263–264
Atomic bomb survivors, 83, 156, 263
Autoradiography, 20–21, 29–31, 35–
    36, 79, 94, 104–105, 138, 168,
    288
Autosome, 19
  aneuploidy, human, 197–198
  nondisjunction of, 149–150
  numerically abnormal, 194–207

Autosome (cont.)
    structurally abnormal, 211–225
    X;autosome translocation, 187–188
Axolotl, 195
5-Azacytidine, 172
Azotospermia, 153, 187

Banded human karyotype, 21–22
Banding studies, on human
        chromosomes, 47
Banding techniques, 1, 3, 15, 53–54
    for cells in culture, 37–38
    of fixed chromosomes, 36–37
    nomenclature of banded
        chromosomes, 38–41
    prophase, 1, 38, 40, 267
Barley, 197
Barr body(ies) (X chromatin), 4, 7, 49,
        166–167, 170–171, 176–177,
        182, 185, 188, 191
    bipartite, 184–186
    formation of, 166–170
B chromosomes, 51–52, 219
beta-Globin gene, 289
Bivalent, 132–133, 136–137, 237
    heteromorphic, 136
Bloom's syndrome (BS), 9, 54, 71, 73,
        88–89, 96–99, 103–108, 259,
        263–264, 270, 285
Bombyx mori, 123
Bone marrow, 33, 96
Bouquet formation, 127–131, 138,
        144–145
BrdU, see Bromodeoxyuridine
Breaks, chromosome, see
        Chromosome breaks
Bromodeoxyuridine (BrdU), 4, 7, 31,
        37–38, 75, 94–95, 138, 169
Buccal smear, 166–168

Cadmium, resistance to, 259
Campomelic dysplasia, 173
Cancer, 8, 48–49, 80–83, 99, 111, 116,
        262–264, 266
    breast, 258
    cervical, 111, 264, 277
    chromosomal development of, 273–
        280

chromosome breakage syndromes
        and, 92
    chromosomes and, 8
    colon, 266, 268
    colon/endometrial, 263
    double minutes and, 256, 258
    fallopian tube, 278
    hepatocellular, 264
    homogeneously stained regions and,
        256
    induction of, 263–264
    with known specific recurrent
        chromosomal defect, 266
    mitotic aberrations in, 276–278
    nasopharyngeal, 264
    predetermination of chromosome
        changes in, 276
    progression of, 273
        effect of chromosome changes on,
        278–280
    small cell lung, 266
Cancer syndrome families, 263
Carbon tetrachloride, 119
Carcinogen, 264, 271
Carcinogenesis, multistep, 271
Carcinoma, see Cancer
Carex, 13
Cat cry syndrome, see Cri du chat
        syndrome
Cat eye syndrome, 204
C-band(ing), 37, 47–52, 82, 96, 212–
        213, 258
Cd-band(ing), 37, 258
Cell culture, 2, 37–38, 83, 262
Cell fusion, 7, 28, 83, 89, 103–107,
        120, 285–286
Cell hybridization studies, 213
Centric banding, see C-band(ing)
Centric fusion, 72–73; see also
        Robertsonian translocation
Centric ring, 66–67
Centriole, 27
Centromere, 13, 133–134
    inactivation of, 71–72, 186, 190, 224
    misdivision of, 72, 151–152, 182,
        244
Centromere index, 15
Centrosome, 27
Chemical theory of chromosome
        breaks, 79–80

Chemotherapy, cancer, 83
Chiasma(ta), 34, 133, 155
    failure of formation of, 149–150
    frequency of, 144
    mitotic, 97–98, 105
    number of, 137–138
    structure of, 136–137
    terminalization of, 134
Chiasma interference, 138
Chiasma-type theory, 136–137
Chimera, 213
Chimpanzee, 4–5, 203
Chinese hamster, 96, 276
Choriocarcinoma, 196, 270
Choroidemia, 171
Chromatid, 14, 29, 58–61
Chromatid breaks, 67–68
Chromatid bridge, 278
Chromatin, 58
Chromocenter, 25–26, 46–50, 53, 143
Chromomere, 47, 61, 115, 128, 131–
    132
Chromonema, 15
Chromosomal imbalance, anomalies
    caused by, 198–200
Chromosomal polymorphisms, 211–
    213
Chromosome
    acentric, 13, 214–215, 228
    acrocentric, 15, 21, 108, 212, 222,
        243–246, 248
    allocyclic, 107–108
    arrangement of, during interphase,
        27–28
    B, 51–52
    banding of, see Banding techniques
    C-minus (CM), 258–260
    condensation of, 121
    dicentric, 66–68, 70, 214–215, 223–
        225, 228
        origin of, 70–71
    differentiation of
        longitudinal, 6–7, 45–55
        molecular, 45–46
    evolution of, see Evolution
    fine structure of, 58–63
    giant, 112
    human, see Human chromosome
    isodicentric, 70–71, 98, 223, 225
    lagging (laggard), 28, 149, 280

    lampbrush, 60–63, 133
    loss of, 28, 149, 207, 286
    marker, 276, 283
    metacentric, 15
    metaphase, see Metaphase
        chromosome
    monocentric, 72, 152, 244–245
    mouse, 286
    multineme, 14
    multiradial, 73–74
    negatively heteropycnotic, 46
    oncogenes and, 262–271
    Philadelphia, 267–268, 275–276
    polytene, see Polytene chromosome
    positively heteropycnotic, 46
    prematurely condensed, see
        Prematurely condensed
        chromosome
    pulverized, 82, 104, 107–108
    ring, 66–67, 69, 94, 108, 219–222
    sex, see Sex chromosome
    shape of, 18
    size of, 17–18
    small extra, 223
    spiralization of, 14, 133, 144
    structure of, 13–24
        abnormal, 211–225
    submetacentric, 15–17
    subtelocentric, 16
    telocentric, 15, 152
    triradial, 73–74
    ultrastructure of, 6
    unineme, 14
    X, see X chromosome
    Y, see Y chromosome
Chromosome bands, 53–54; see also
        Banding techniques
    human, function of, 54–55
Chromosome breakage syndromes, 9–
        10, 88–92, 263–264
    cancer and, 92
Chromosome breaks, 7, 9, 24, 65–68,
        211
    causes of, 65–66, 78–85
        genetic, 82, 88–92
        chemically induced, 79–81
        fragile regions and, 74–75
        at hot spots, 82
        methods in study of, 83–85
        multiple, 68, 230

Chromosome breaks (*cont.*)
  nonrandomness of, 82–83
  radiation-induced, 79–80
  reciprocal translocations and, 229–230
  spontaneous, 78–79
  virus-induced, 80–82
Chromosome changes
  aneuploid, 23
  euploid, 21–23
Chromosome number, 1, 17
  human, abnormalities of, 194–207
Chromosome rearrangements, 66–68
Chromosome replication, 25–31
  saltatory, 260
Chrysanthemum, 23
Clastogenic substance, 80
Clinical cytogenetics, 10
Clone panel, 286
C-mitosis, 112, 116, 278
Colchicine, 2, 13, 32, 34, 116, 154, 274, 288
  resistance to, 259
Collagen gene, 290
Color blindness, 283
Constriction
  primary, 13–15
  secondary, 15, 48, 51
Coriphosphine O, 38, 95, 170
Corn, 197
Cri du chat syndrome, 69, 198, 216–219, 235
Crossing-over, 6, 52, 125–127, 132–141, 147, 214–215, 222, 249
  mitotic, 89, 97–98, 125, 270
  regulation of, 52
  unequal, 100, 259–260
Crown gall, 262–263
Cryptorchidism, 199
Cystadenocarcinoma, papillary, of ovary, 266
Cytochalasin B, 288
Cytogenetics
  human, *see* Human cytogenetics
  somatic cell, *see* Somatic cell cytogenetics

DAPI stain, 37, 52, 212
Datura, 197

Deletion, 65, 69, 214–216, 219–222, 232–233, 270
  5p−, *see* Cri du chat syndrome
  in gene mapping, 287
  interstitial, 67
  partial, of X chromosome, 183
Deoxyribonucleic acid (DNA), 6, 14
  content of nucleus, 18–19, 29, 41
  highly repeated sequence, 45
  methylation of, 172
  middle repeated sequence, 45, 46
  packaging in chromatids, 58–61
  in polytene and lampbrush chromosomes, 61–63
  recombinant, 290–291
  repeated sequences of, 45
  replication of, 29–31
  rings, 63
  satellite, 50, 288
  simple sequence-repeated, 50
  single-stranded probe of, 288
  synthesis of, 29, 126–130, 134–135, 143
  unique sequence, 45–46
Diakinesis, 130, 133
Dictyotene, 143
Differentiation, 9, 120
  of chromosome
    longitudinal, 45–55
    molecular, 45–46
  sexual, 162
  somatic cell cytogenetics and, 123–124
Dihydrofolate reductase, 259
7-12-Dimethylbenzα anthracene (DMBA), 276
Diphtheria toxin, 286
DIPI stain, 37, 212
Diplochromosome, 29–31, 98, 112, 217–218
Diploid generation, 125
Diplonema, *see* Diplotene
Diplotene, 129–130, 133–135, 139, 143–144
Dispermy, 157
Distamycin A, 75
"Distance pairing" stage, 128–129, 138, 143
DM, *see* Double minute
DNA, *see* Deoxyribonucleic acid

Domain, 61
Donkey, 171
Dosage compensation, 4, 170
Double minute (DM), 100, 256–257,
    268, 276
  as expression of gene amplification,
    259
  origin of, 260
  relationship to homogeneously
    stained region, 259–260
  significance of, 260
  structure of, 256–258
Down's syndrome (mongolism, 21
    trisomy), 3, 21, 55, 141, 156,
    194, 203, 221–222, 245–252
Drosophila, 1, 19, 46, 61, 80, 105,
    114–115, 125–126, 139, 154,
    156, 161, 172, 178, 228, 231,
    252, 289, 291
Duchenne muscular dystrophy, 287
Duffy blood group locus, 283
Duplication, 100, 214–215, 222–223,
    232–233

Edward's syndrome, see Trisomy, 18
Egg, 142; see also Oocyte
  "empty", 196
Embryo sac, 142
Endocycle, 110
Endomitosis, 110, 112, 115–116, 121,
    196, 278
Endopolyploidy, 120, 123
Endoreduplication, 73–74, 110–112,
    120–121, 157, 196, 278
Epicanthic folds, 217
Epstein-Barr virus, 271
Eremurus, 25, 126–134, 143–144
Erythroblastosis virus, 265
Euchromatin, 46–47
  underreplication of, 122–123
Eukaryote, 13
Euploid chromosome changes, 21–23
Evolution
  centromere inactivation and, 224
  of chromosome number, 17
  of chromosome shape, 18
  crossing-over and, 126
  gene amplification and, 101, 261
  of human chromosomes, 4–5, 47

  of karyotypes, 21–23
  polyploidy and, 194–195
Exclusion mapping, 288

Family studies, 283
Fanconi's anemia (FA), 73, 88–91, 96,
    263
Fertilization, 125
Fetal death, 223–234
Feulgen squash technique, 32, 34, 126,
    129, 131, 132
Fibroblast, 33, 78, 111, 166–171, 260
Fission, of centromere, see
    Misdivision, of centromere
Flow cytometry, 41, 120
Fragile region, 7, 74–75, 186, 190–
    191, 213–214, 264, 283
Fritillaria, 19, 51, 138

Gamete, 142
  diploid, origin of, 157
Gap, in chromosome, 67, 84–85
Gap 1 (G_1) phase, 29
Gap 2 (G_2) phase, 29
G-band(ing) (Giemsa-band), 36–39,
    47–48, 51, 132, 257–258
Gene
  active, 54–55
  human, number of, 291
  localization of, 36, 289–290
  syntenic, 283
  X-linked, 283, 287, 291
  Y-linked, 283
Gene amplification, 100–101, 122–
    123, 256, 260–162, 268–269
Gene assignment
  confirmed, 283
  controversial, 283
  provisional, 283
Gene dosage, 288
Gene mapping, 105, 213, 282–283; see
    also Mapping of human
    chromosomes
Genetic counseling, 10, 231
Gerris, 115–116
Giemsa stain, 36, 37–38, 95, 169–170
Glucose-6-phosphate dehydrogenase
    (G6PD), 171
  gene for, 287

Goat, 172
Gonadal dysgenesis, 172, 185, 187,
    231
Gorilla, 5, 203
Graminae, 23
Growth hormone gene, 290
Growth retardation, 199

Hair-root follicle (cell), 35, 168
Hamster egg, fertilization with human
    sperm, 34, 152, 252
Haploid generation, 125
Haploid number, 17
Haploidy, 21, 194
Haplopappus, 16-17, 48
HeLa cells, 274
Hemangiomata, 201-202
Hermaphroditism, 173, 180
Heterochromatin, 6, 18, 46
    constitutive, 37, 45, 47, 50-52, 212-
        213, 291
        breakage of, 47
    facultative, 50, 52-53
    intercalary, 53
    variation in, 100
Heterokaryon, 285
Heteromorphisms, see
    Polymorphisms
High-resolution banding, see Banding
    techniques, prophase
Histones, 15, 58, 61, 289
Hodgkin's disease, 264
Hoechst 33258, 37-38, 169-170
Holoprosencephaly, 198, 201
Homogeneously stained region (HSR),
    256-261, 268, 278
    as expression of gene amplification,
        259
    origin of, 260
    relationship to double minute, 259-
        260
    significance of, 260-261
Horse, 171
Hot spot, 82, 98, 232, 270
HSR, see Homogeneously stained
    region
Human chromosome complement,
    19-21

Human chromosomes
    banding studies on, 47
    lengths of, 19-21
    mapping of, see Mapping of human
        chromosomes
    nomenclature of, 5, 38-41
    replication of, 29-30
Human cytogenetics
    clinical, 10
    future of, 5-10
    methods in, 32-41; see also
        Banding techniques
    autoradiography, 35-36, 168, 170
    direct, 32-33
    meiotic studies, 34-35
    prenatal, 33-34
    quantitative, 41
    tissue culture, 33
    past of, 1-4
H-Y antigen, 4, 7, 162-163, 172-173,
    283
H-Y antigen regulator, 162-163
Hybridization
    donkey/horse, 171
    plant, 23, 82
    in situ, 36, 49, 265, 288-290
    somatic cell, 265, 284-286
Hybridoma, 105
Hydatidiform mole, 115, 120-121,
    196, 213, 270
Hypertelorism, 216
Hypodiploid, 112
Hypotonia, 216
Hypoxanthine-guanine
    phosphoribosyl-transferase
    gene (HGPRT), 289

Idiogram, 19
Immunodeficiency, 74, 89, 91-92
Immunoglobulin gene, 290
Infertility, male, 153, 231, 243, 248
Insertion, 222
in situ hybridization, see
    Hybridization, in situ
Insulin gene, 290
Interchromosomal effects, 252-253
Interferon gene, 290
Interkinesis, 132, 134

Interphase, 25–26, 126, 128, 130
  chromosome arrangement during,
    27–28
  premeiotic, 128, 130, 135
Interphase nucleus, 9, 25–26, 46, 49
  structure of, 7
Intersexuality, 172–173
Inversion, 131, 213, 252
  paracentric, 67, 215
  pericentric, 66–68, 213–215
Ionizing radiation, see Radiation
Isochromosome, 70–71, 98, 151–152,
    182–185, 222–223, 244

Juncus, 13

Karyotype, 19
  of balanced carrier of t(5p−;12q+),
    218
  banded human, 21–22, 39
  of 8-trisomic cell, 205
  human, 2
    normal female, 22
    normal male, 18
  of 13-trisomic male, 200
  partial, 224, 245
Kinetochore, see Spindle attachment
Klinefelter's syndrome (47, XXY), 3,
    153, 173, 176–177, 179

Lampbrush chromosome, 60–63, 133
Leptonema, see Leptotene
Leptotene, 126–130, 135, 138, 143
Leukemia, 203, 256, 263–267, 274–
    276, 280
  acute lymphocytic, 266
  acute nonlymphocytic, 266
  chronic granulocytic, 266
  chronic lymphocytic, 266
  chronic myelogenous, 267–268,
    275–276
Lilium, 128
Linker, deoxyribonucleic acid, 58–59
Linum usitatissimum, 19
Liver, 119–120
Lotus tenuis, 17

Louis-Bar syndrome, see Ataxia
  telangiectasia
Lymphocyte, 3, 14, 33, 95, 168
  cultured human, mitotic cycle in, 26
Lymphoma, 91, 265–267, 276
  Burkitt's, 264, 266, 268, 271
  small cell, 267
Lyon hypothesis, 4, 170–171

Maize (corn), 156, 197
Major histocompatibility complex
    (HLA), 283–284
Malignant melanoma, 230
Mapping of human chromosomes, 8,
    282–291
  using cell hybridization, 284–286
  using deletions, 287
  using family studies, 283
  using gene dosage, 288
  using in-situ hybridization, 288, 290
  using marker chromosomes, 283–
    284
  using microcells, 288
  using recombinant DNA, 290–291
  using single chromosomes, 288
  using translocations, 286–287
Measles, 82
Megakaryocyte, 120
Meiosis, 4, 6, 125–135, 136–147
  abnormal, 149–157
  cytologic studies of, 34–35
  human, 34–35, 142–147
  sex chromosomes in, 172
  significance of, 125–127
  stages of, 126–135
  in translocation carriers, 237–238,
    252
Melandrium, 161
Meningioma, 69, 266, 269
Mental retardation, 75, 177–179, 188–
    191, 198–202, 212, 216–221,
    230–234, 283
Metaphase, 26–27, 126, 144–145
  in cancer cells, 277–279
  multipolar, 116, 277–279
Metaphase chromosomes, 13–16
  characterization of, 15
  spiralization of, 14–15

Metaphase I, 130, 132–134
Metaphase II, 130, 132, 134
Metaphase plate, 27–28, 133, 257
Methotrexate resistance, 256, 259
1-Methyl-2-benzylhydrazine, 14–15, 80–81
Micrasterias, 123
Microcell, 290
Microcephaly, 216
Microchromosome, 18
Micronucleus, 28, 84, 107–108, 112, 116, 258
Microphthalmia, 201–202
Minute, 67
  double, *see* Double minute
Misdivision, of centromere, 72, 151–152, 182, 244
Mitomycin C, 83, 96, 248
Mitosis
  in cancer cells, 262, 276–278
  modifications of, 110–117
    C-mitosis, 112, 116, 278
    endomitosis, 110, 112, 115–116, 121, 196, 278
    endoreduplication, 73–74, 110–112, 120–121, 157, 196, 278
    multipolar mitosis, 112, 116–117, 276–279
    polytenization, 112–115
    restitution, 112, 116, 154, 157, 195
Mitotic chiasma(ta), 67, 68–69, 74, 97–99
  nonrandom localization of, 98–99
  origin of, 100
Mitotic crossing-over, 10, 89, 97–98, 125, 270
Mitotic cycle, 25–31
  length of, 28–29
Mitotic recombination, segregation after, 99
Mongolism, *see* Down's syndrome
Monosomy, 10, 23, 28, 194, 199, 206–207, 221, 279
  partial, 215–216, 231–233, 235
  tertiary, 235
  21, 194
Monosomy 21 syndrome, 206

Mosaicism, 28, 71, 162, 170, 180, 196–204, 213, 221–225, 234, 245
  X chromosome, 180, 184
Mouse, 15, 18, 80–81, 111, 138, 144, 153, 171–173, 228, 231, 248, 256, 263, 274. 276, 284–286
Multigene family, 45–46
Multiple endocrine adenomatosis, 92
Multivalent, 138
Mutagen, 82–84, 96–97, 264
Mutagenesis, 9, 65
  studies using sister chromatid exchanges, 96–97
Mutation, 263
  chemically induced, 96
  chromosome, 21, 23
  gene, 21
Myelocytomatosis virus, 265
Myeloblastosis virus, 265
Myocardium, 120
Myosin heavy-chain gene, 290

Neuroblastoma, 256–260, 266, 271
Newborn
  chromosome abnormalities in, 238
  Robertsonian translocation in, 246–247
Newt, 195
Nomenclature, of human chromosomes, 5, 38–41
Nondisjunction, 28, 252
  of autosomes, 149–150
  genetic causes of, 156
  maternal age and, 154–156
  meiotic, environmental causes of, 153–154
  paternal age and, 155
  secondary, 141
  of sex chromosomes, 150–151
Nucleolar material, 258
Nucleolar organizer, 15, 26, 48, 69
Nucleolus(i), 26, 48–50, 63, 143
Nucleoside phosphorylase gene (NP), 287
Nucleosome, 58–59, 61
Nucleosome core, 58

Nucleus
  deoxyribonucleic acid content of,
    18–19, 29, 41
  interphase, see Interphase nucleus

Ocular albinism, 171
Oligomenorrhea, 187
Oligospermia, 153, 187, 248
Oncogene, 8, 231, 264–265
  amplification of, 260
  chromosome and, 262–271
  examples of, 265
  recessive, 270–271
  reciprocal translocations and 265–
    269
Onion, 50, 120, 122–123
Oocyte
  human, meiosis in, 142–144
  human embryonic, 127
Ophioglossum reticulatum, 17
Orangutan, 5, 203
Ouabain, 286

Pachynema, see Pachytene
Pachytene, 127–132, 135, 141, 144–
    145, 236–237
Paris, 46, 138
Parthenogenesis, 23
Patau's syndrome, see Trisomy, 13
Paternity dispute (determination),
    212–213
PCC, see Prematurely condensed
    chromosome
Pea root, cultured, 113, 115
Pentaploidy, 195
Peromyscus, 50
Persistence of fetal hemoglobin, 198
Phaseolus, 115
Philadelphia chromosome, 267–268,
    277–278
Philaemus, 139
Phosphoglycerate kinase gene (PGK),
    287
Phosphonacetyl-L-aspartate, (PALA),
    resistance to, 259
Phytohemagglutinin, 3, 33

Pinus resinosa, 51
Placenta, 120, 195, 277
Pleurodeles waltlii, 60, 62
Polar body, 142, 178, 195
Polycomplex, 139
Polydactyly, 198, 202
Polyethylene glycol, 285
Polymorphisms, chromosomal, 211–
    213
Polymorphonuclear white blood cells,
    166, 170
Polyploidy, 21, 23, 111–112, 122–124
  human, 194–195
  meiosis and, 141–142
  somatic, 120–121
Polytene chromosome, 61–63, 112–
    115, 122, 289
Polyteny, 112–115
Position effect, 8, 187, 189, 231, 268
Prader-Willi syndrome, 216, 287
Prefixation treatment, 2, 33
Prematurely condensed chromosome
    (PCC), 103–105, 120
  formation of, uses of, 105–107
Prenatal cytogenetic studies, 33–34
Prochromosome, 46, 128
Prokaryote, 13
Prometaphase, 26–27, 133
Prophase, 26–28, 126, 135, 142
  in cancer cells, 277–278
Prophase banding, 38, 216, 283
Protein, chromosomal, 54; see also
    Histones
Proto-oncogene, 8, 264–265
Pseudogene, 46
Puff, chromosome, 61–63, 115

Q-band(ing), 36–39, 47–48, 53–54, 98,
    160–161, 200, 205, 212–213,217
  base ratio of, 54
Quadriradial, 67–72, 91
Quadrivalent, 141
Quinacrine, 36, 53

Rabbit, 276
Rabl orientation, 27, 107

Radiation, 65, 79, 83, 97, 154–156,
    248, 263–264
Radiomimetic substance, 80
Rat, 172, 274, 276
R-band(ing), 37, 39, 47–48, 212, 217
Reciprocal translocations, 8, 66, 68,
    79, 215, 228–240, 283; see also
    Translocation carriers
  breakpoints in, 229–230
  examples of translocation families,
    234–237
  multiple rearrangements in, 230
  occurrence of, 228
  oncogenes and, 265–269
Recombination nodule, 6, 141, 147
Reindeer, 16–17
Reproduction, asexual, 23
Restitution, 112, 116, 154, 157, 195
Restriction endonuclease, 290
Retinoblastoma, 216, 266–267, 270,
    287
Retrovirus, 264
Reverse banding, see R-band(ing)
Ribonucleic acid (RNA), 61–63
  synthesis of, 115, 123
Ribosomal RNA genes, 48, 99, 123,
    212, 259–260, 289
Ring chromosome, 66–67, 69, 94, 108,
    219–222
RNA, see Ribonucleic acid
RNA polymerase, 62–63
Robertsonian translocation, 70–73,
    153, 203, 231, 243–253
  ascertainment of
    through infertility, 247–248
    through unbalanced individual,
    247
  balanced, 243
  dicentric, 244–245
  frequencies of types of, 246
  monocentric, 244–245
  in newborns, 238, 247
  nonrandomness of, 248–250
  occurrence of, 243–244
  segregation in carriers of, 250–
    252
Roberts' syndrome, 92
Rosaceae, 23

Sarcoma
  Ewing, 266
  mouse, 256
  SEWA mouse, 258–260
Satellite, 15, 19, 48, 248
Satellite stalk, 15, 48, 99–100, 212,
    248, 259–260, 289
Scilla, 133
Semiconservative replication, 29–31
Sendai virus, 285
Sex chromatin, see Barr body
Sex chromosomes, 7; see also X
    chromosome; Y chromosome
  abnormal, male sterility and, 153
  human, 4, 19
  nondisjunction of, 150–151
  numerical abnormalities of, 176–
    180
Sex determination, 7, 23
  in man, 1, 3, 160–164
Sex reversal, 172–173
"Sex vesicle", 144–145
Sézary's syndrome, 92
Silver staining
  NOR, 37, 48–49
  Synaptonemal complex, 34, 140–
    141
Sister chromatid, 14
Sister chromatid exchange, 4, 9, 38,
    84, 89–91, 220, 222
  biological significance of, 97
  detection of, 95
  in mutagenesis research, 96–97
  occurrence of, 94–96
  unequal, 97, 100, 222, 260
Somatic cell cytogenetics, 8–9, 119–
    124
  differentiation and, 123–124
  history of, 119
  methods of, 119–120
Somatic polyploidy, 120–121
Spacer, 46
Spectrocytophotometry, 41
Sperm
  diploid, 157
  human
    chromosomally abnormal, 152
    infertility and, 153

Spermatid, 142
Spermatocyte, human, meiosis in, 144–147
Spinach, 120, 197
Spindle, multipolar, 112, 116–117, 154
Spindle, 27
Spindle attachment, 13
Stemline, 274
Sterility, see Infertility
Steroid sulfatase gene (STS), 162
Stillbirth, 233, 238–240
Strabismus, 217
Swyer's syndrome, 172–173
Synapsis, 131
Synaptonemal complex, 6, 34, 138–141, 147, 154
Synkaryon, 285
Syrian hamster, 140–141

Target theory of chromosome breaks, 79–80
T-band(ing), 37
Telomere, 68–70, 225, 229
  association between, 70
Telophase, 26–29, 126, 130, 134
Temperature, effect on meiosis, 154
Testicular feminization syndrome (Tfm), 173
Tetraploidy, 21, 111–112, 154, 157, 194–195
  human, 195–197
Tetrasomy, 23
  partial, 223
Thymidine, tritiated, 29–31, 35, 79, 94, 104–105, 170
Thymidine kinase gene (TK), 286–287
Tissue culture, 33
Tooth size, 164, 173
Tradescantia, 133
Transcription, 62–63
Transfection, 265
Translocation(s), 10, 153
  Cattanach, 171
  in gene mapping, 286–287
  interchromosomal effects of, 252–253

involving Y chromosome, 188–189
in newborn, 238
reciprocal, see Reciprocal translocation
Robertsonian, see Robertsonian translocation
unbalanced, 236–237
X;autosome, 187–188
Translocation carriers
  balanced, phenotypes of, 230–231
  examples of translocation families, 234–237
  fetal death of offspring of, 233–234
  genetic risks for, 238–240
  meiosis in, 237–238
  Robertsonian, 250–252
  unbalanced, phenotypes of, 231–233
Transposable elements, 45–46
Trillium, 46, 128
Triploidy, 21, 195
  human, 195–196, 207
Trisomy (trisomic), 23, 28, 54–55, 156, 195–198, 269, 279
  8, 199, 204–206
  18 (Edward's syndrome), 3, 154, 156, 194, 197–199, 203–204
  11q, 55
  15, 204, 206
  14, 204, 206
  interchange, 237, 239
  9, 204, 206
  19, 204, 206
  partial, 10, 229, 231–233, 283
  pure partial, 222–223
  7, 204–206
  16, 204, 206–207
  10, 204, 206
  tertiary, 235, 239
  13, (D₁, Patau's syndrome), 3, 152, 154, 194, 197–203, 207, 247, 250–251
  partial, 203
  symptoms of, 201–202
  3, 204–205
  12, 204, 206
  20, 199, 204, 206

Trisomy (trisomic) (*cont.*)
21 (Down's syndrome, mongolism),
55, 141, 154, 156, 194, 197–
199, 203, 250–252
22, 3, 204
Trisomy mapping, 288
Trisomy period, 2–3
Triticale, 23
Triturus vulgaris, 127
Trivalent, 141–142
Trophoblast, 196
Trophoblast cell, 115, 123
Trophoblastic villus biopsy, 34
Tumor
ascites, 274
benign, 262, 266, 269, 280
chromosome studies in, 274–276
malignant, 120–121, 262; *see also*
Cancer
parotid gland, 266, 269
solid, 256, 267, 271
Turner's syndrome, 3, 154, 161–162,
177, 180, 183–185
symptoms of, 177–178
Two-plane theory, 136–137
Tyrosine kinase, 264

Ultraviolet laser microbeam
technique, 7, 28
Ultraviolet light, 79, 96
Underreplication, 122–123
Univalent, 149–150, 155
Uterus, bicornate, 199
U-type exchange, 138, 249–250

Vaginal smear, 168
Ventricular septal defect, 199
Vicia, 17, 52, 65, 82
Vincristine, resistance to, 259
Virus
induction of cancer by, 264–265
induction of chromosome breaks
by, 80–82
Rous sarcoma (RSV), 264–265, 276

Werner's syndrome, 92
Wilms' tumor, 216, 266–267, 271, 287

X chromatin, *see* Barr body
X chromosome, 7, 19–20
abnormal,
consisting of X material, 182–185
female phenotype and, 189–190
structurally, 182–192
always-active region on, 185, 188–
189
aneuploidy of
with female phenotype, 176–178
with male phenotype, 177, 179
behavioral abnormalities of, 190–
191
critical region on, 7, 187–189, 231
cytogenetic diagram of, 185
dicentric, 190
evolution of, 5
fragile region on, 75, 186, 190
human, 166–173
inactivation center on, 7, 184–186
inactivation of, 7, 170–171, 190–
192
inactive, 4, 52–53, 166–168
late-labeling, 38
late-replicating, 38, 52, 171
during meiosis, 143–147
misdivision of, 151
nondisjunction of, 150–151
partial deletion of, 183–185
reactivation of, 172
ring, 183
single active, 4, 170
structure of, 166
telocentric, 183, 185
X;autosome translocation, 187–188
Xeroderma pigmentosum, 92
Xg blood group, 162, 176
X-rays,, *see* Radiation
XX male, 173
XXX female, 3, 166–169, 176–178
XXXX female, 177–178
XXXXX female, 177
XXXXY male, 3, 163, 177, 179
XXXY male, 177
XXXYY male, 177, 179
XXY male, 153, 161, 177, 179
XXYY male, 177, 179
XXYYY male, 177
XY body, 144–145

XY gonadal dysgenesis, 172-173
XYY male, 153, 161, 177, 179
XYYY male, 177, 179
XYYYY male, 176, 179

Y-body, 160-161
Y chromosome, 3-4, 7, 160-164, 176
  abnormal, 162, 164
  aneuploidy of, 177, 179
  dicentric, 162, 164
  genes on, 163-164

human, 160-164
  cytogenetic diagram of, 163
  during meiosis, 144-147
  polymorphism, 212
  ring, 164
  translocations involving, 188-189

Zygonema, see Zygotene
Zygote, 125
Zygotene, 128-131, 135, 138, 141-
  144

Printed in the United States
By Bookmasters